THE AQUASCAPING MASTER COLLECTION

The Step-by-Step Handbook to Designing and Maintaining Freshwater Aquariums.

Avoid Costly Mistakes, Master Plant and Fish Care, and Create Thriving Aquatic Habits

Benjamin Thorne

TABLE OF CONTENTS

INTRODUCTION

Aquascaping is more than just creating an attractive underwater landscape—it's a combination of art, science, and nature working in harmony. Imagine your living room transformed into a serene ecosystem where every plant, rock, and fish plays a part in maintaining balance and beauty. Whether you're new to the hobby or an experienced aquascaper, you're about to dive into a world where creativity meets biology. Welcome to The Aquascaping Master Collection: Freshwater Aquariums, where we explore everything you need to know about creating and maintaining stunning freshwater aquascapes.

Welcome to Aquascaping

If you've ever gazed into a beautifully maintained aquarium and felt a sense of peace, you've already experienced the power of aquascaping. It's the art of arranging aquatic plants, stones, and driftwood in a visually pleasing and balanced way inside an aquarium. But aquascaping goes beyond aesthetics; it fosters a self-sustaining ecosystem that benefits both its inhabitants and the hobbyist.

Aquascaping offers a unique opportunity to design your own aquatic world, turning your aquarium into a masterpiece. As you delve deeper into the practice, you'll find that aquascaping is a form of personal expression, much like painting or gardening. The goal is not only to create something visually appealing but also to ensure that it thrives as a healthy ecosystem.

What is Aquascaping?

Aquascaping can be described as underwater gardening, but it is much more than that. At its core, aquascaping is the craft of arranging natural elements like rocks, plants, and wood within an aquarium to form an aesthetically pleasing and sustainable underwater environment. This discipline demands attention to detail, patience, and an understanding of how various elements interact with each other to create a balanced ecosystem.

The roots of modern aquascaping trace back to the early 1990s, when Takashi Amano, a Japanese photographer and aquarist, introduced the Nature Aquarium style. He demonstrated how one could create beautiful aquascapes that mimic the tranquility of nature using plants, rocks, and driftwood. Today, there are numerous styles of aquascaping, from the minimalist Iwagumi layouts to lush Dutch-style aquascapes. Each style has its own principles, but all share the common goal of achieving harmony between the aquarium's inhabitants and their environment.

The Benefits of Aquascaping

Aquascaping offers far more than just the opportunity to create a beautiful aquarium. It's a hobby that touches on multiple aspects of well-being, blending artistic expression with therapeutic value and environmental awareness. By understanding these benefits, you can appreciate how aquascaping enriches both your space and your life.

Aesthetic Appeal

At its core, aquascaping is an art form. You're not just setting up an aquarium—you're crafting an underwater landscape that brings nature's beauty into your home. Whether you want to recreate a lush tropical jungle, a tranquil riverbed, or even mimic the appearance of a serene forest floor, aquascaping allows for endless creativity. The sight of a vibrant, well-maintained aquascape can transform any room, making it not only a point of pride but also a focal point for conversation.

For instance, adding vibrant plants like Vallisneria, striking stones, and carefully selected fish can turn a simple tank into a tropical paradise. People often remark on how aquariums bring a sense of calm and wonder to a space, whether it's a living room, office, or even a waiting area. Studies even suggest that aquariums are associated with greater workplace satisfaction and productivity. The ability to shape your aquarium to reflect a particular style or environment gives you creative control over your space, making it a dynamic piece of living art.

Therapeutic Benefits

Beyond its visual appeal, aquascaping offers significant therapeutic benefits. Many aquascapers find that the hobby provides a calming escape from the hustle of daily life. The act of designing, planting, and maintaining an aquarium can be meditative, helping to reduce stress and anxiety. There's something uniquely satisfying about nurturing a thriving, balanced ecosystem from scratch, where every element—from the plants to the fish—plays its part.

Research backs this up. Studies have shown that watching fish swim in a well-maintained aquarium can lower blood pressure, reduce anxiety, and even improve mental health. Hospitals, nursing homes, and therapy offices often feature aquariums for this very reason—they promote a sense of calm and well-being. Additionally, aquascaping requires attention to detail and patience, offering a rewarding way to disconnect from the fast-paced demands of modern life.

Environmental Impact

Aquascaping isn't just about what happens inside your tank—it can also have a broader positive impact on the environment. As aquascaping grows in popularity, there has been a shift toward using sustainably sourced plants, fish, and materials. This emphasis on sustainability helps reduce the strain on wild aquatic ecosystems. Many hobbyists now choose plants and fish that are grown in captivity or sourced from responsible breeders rather than those taken from the wild.

Moreover, creating a self-sustaining ecosystem within your aquarium encourages environmentally friendly practices. Aquascapes that thrive with minimal external intervention, like chemicals or excessive water changes, help reduce the overall environmental footprint of the hobby. Additionally, aquascaping can foster a deeper appreciation for aquatic ecosystems, encouraging hobbyists to become more aware of the importance of maintaining balance in both natural and artificial environments. This awareness often translates into more conscious decisions about how we interact with nature, whether through supporting conservation efforts or adopting more sustainable lifestyles.

Overview of Freshwater Aquascaping Styles and Trends

The world of freshwater aquascaping offers diverse styles and evolving trends that cater to various tastes. The Dutch style, one of the oldest, is characterized by densely planted tanks with layered, colorful, and textured plants arranged in terraces to create depth. In contrast, the minimalist Iwagumi style focuses on rocks as the main feature, using a few plants to create a harmonious, serene environment inspired by Japanese Zen gardens.

Biotope aquascaping aims to replicate specific ecosystems, using native plants, fish, and substrates to mimic natural habitats like Amazonian riverbeds. This trend emphasizes ecological accuracy and biodiversity. The jungle style offers a more untamed aesthetic, focusing on dense plant growth and creating a wild, overgrown look reminiscent of tropical forests.

Modern trends include minimalism and naturalism, where aquascapers favor serene, uncluttered designs using fewer plants and hardscape materials to evoke balance. Biotope authenticity remains popular as aquarists strive to mirror real-world habitats. Low-tech setups, emphasizing sustainability and reduced maintenance, appeal to environmentally conscious hobbyists, while advances in LED lighting allow for customizable, dynamic displays that enhance the visual experience.

Aquascaping has also become a tool for well-being, with designs focused on relaxation and mindfulness. These tranquil aquariums often feature soothing water flow, soft lighting, and harmonious layouts. Overall, aquascaping blends artistic expression, environmental consciousness, and personal wellness, offering endless possibilities for enthusiasts to explore.

Who This Book Is For: Beginners to Advanced Enthusiasts

Whether you're just dipping your toes into aquascaping or you've been creating stunning aquascapes for years, this book is designed for you. Beginners will find detailed, step-by-step instructions on how to set up their first aquascape, while more experienced aquascapers can explore advanced topics like CO2 injection, substrate choices, and aquascaping techniques used in international competitions.

For the novice, aquascaping might seem overwhelming at first—there's equipment to buy, plants to choose from, and an ecosystem to balance. This book breaks everything down into manageable steps so you can confidently embark on your aquascaping journey.

For advanced enthusiasts, this book delves into the intricacies of creating more complex aquascapes, offering expert tips and tricks on plant care, water chemistry, and design principles that can elevate your aquascapes to the next level. It's a resource that you can return to time and again whether you need a refresher on maintenance techniques or are looking for inspiration for your next project.

How to Use This Book Effectively

As with any creative endeavor, planning is key when it comes to aquascaping. This book is structured to guide you through the process from start to finish, helping you navigate everything from choosing the right equipment to mastering the art of aquascape design.

Start by reading the foundational chapters that introduce the core concepts of aquascaping. These sections cover everything from tank setup and plant selection to understanding the balance between fish, plants, and water chemistry. Once you have a solid grasp of the basics, you can move on to more advanced topics such as aquascaping techniques, CO2 systems, and troubleshooting common issues.

Throughout the book, you'll find actionable tips, expert advice, and step-by-step guides that will help you turn your vision into reality. Don't be afraid to experiment and push the boundaries of your creativity. Whether you're creating a simple aquascape or a more elaborate layout, the principles outlined in this book will provide you with the tools to succeed.

As you work through each chapter, take your time and enjoy the process. Aquascaping is not a race—it's an ongoing journey where every step offers a chance to learn, grow, and improve. Let this book be your companion and guide as you embark on the rewarding adventure of aquascaping.

By the end of this book, you'll not only have the knowledge and confidence to create stunning aquascapes, but you'll also develop a deeper connection with the aquatic world. The journey to becoming an aquascaping master begins here, so let's dive in and start creating!

CHAPTER

01

GETTING **STARTED** WITH **AQUASCAPING**

"Aquascaping is not just about plants or rocks or driftwood;

it is about creating a world that feels natural and makes you forget it's an aquarium."

Takashi Amano

Imagine a lush, underwater world teeming with life, where vibrant plants sway with the current and colorful fish dart between rocks and roots. Aquascaping is much more than just a hobby; it's an art form. This dynamic and visually stunning practice allows you to craft a slice of nature right in your home, blending the principles of landscape design with the serenity of an aquatic environment. Whether you're a seasoned aquarist or just getting your feet wet, aquascaping offers a unique way to combine creativity and care for living ecosystems. And the best part? It's not just about beauty — it's about creating balance, fostering life, and learning the delicate art of nature.

You might be wondering, why aquascaping? What makes it so important? For many, it's the therapeutic process of designing and maintaining an aquarium. For others, it's the pride in nurturing an ecosystem where plants, fish, and bacteria coexist harmoniously. If you're passionate about nature and have an eye for design, aquascaping offers the perfect opportunity to explore both. But before you dive in, it's crucial to understand that successful aquascaping goes beyond simply arranging rocks and planting greenery. It requires research, planning, and a deep understanding of how these underwater ecosystems function.

In this chapter, we'll take you through the foundational elements of aquascaping, from its fascinating history and evolution to the core concepts you need to grasp before starting your own underwater garden. We'll also explore the importance of research and planning, and how to achieve the perfect balance between plants, fish, and bacteria — the trio that keeps your aquarium thriving.

The History and Evolution of Aquascaping

Aquascaping, at first glance, might seem like a modern innovation, but it actually has deep historical roots. The practice of designing and maintaining aquatic environments has evolved over centuries, from its humble beginnings to the intricate and artistic discipline we know today. By understanding the history of aquascaping, you can better appreciate the art form's depth and how it connects to broader ideas about nature, balance, and creativity.

Aquascaping's origins stretch back to ancient China, where goldfish were kept in ponds as early as 500 AD. This early form of aquascaping wasn't about aesthetics but survival and sustenance. The Chinese built these ponds not just for display but to ensure the health of the fish in their care. It was a utilitarian practice that later spread to Japan and other parts of Asia. However, these early aquascapers were laying the groundwork for something much bigger, even if they didn't realize it at the time.

In the 19th century, aquascaping began to take on a more decorative role with the introduction of glass aquariums in Victorian England. For the first time, people could bring aquatic life into their homes in a way that allowed for both viewing and interaction. Victorians were fascinated by exotic fish and aquatic plants, and the aquarium hobby took off, becoming a popular pastime. However, it was still far from the artistic expression we associate with aquascaping today. These early aquarium enthusiasts were more focused on the novelty of keeping fish than creating an underwater landscape.

The real shift in aquascaping came in the mid-20th century when advances in aquarium technology allowed for more stable aquatic environments. With better filtration systems, lighting, and water quality management, aquarists were no longer just trying to keep fish alive; they had the tools to create beautiful, thriving ecosystems. This set the stage for a new wave of aquascaping, one that embraced both functionality and design.

One of the most influential figures in modern aquascaping is Takashi Amano, a Japanese photographer and aquascaper who revolutionized the field in the 1990s. Amano introduced the concept of "nature aquariums," which focused on mimicking natural landscapes within the aquarium. His philosophy was simple but profound: let nature be the artist. Amano believed that aquascapes should reflect the organic beauty found in nature — the way rivers flow around rocks, the way moss clings to stones, and the way plants grow in harmony with their surroundings.

Amano's work transformed aquascaping into a true art form. He emphasized balance, natural rhythms, and the importance of subtlety in design. Instead of cramming as many fish or plants into an aquarium as possible, Amano taught that the best aquascapes are those that create a sense of peace and harmony, much like the natural environments they aim to replicate.

Amano's influence spread rapidly, particularly in Europe and North America, where his ideas resonated with aquascapers looking to push the boundaries of what could be done within the confines of an aquarium. Competitions began to spring up, with enthusiasts from around the world showcasing their designs. The International Aquatic Plants Layout Contest (IAPLC), one of the most prestigious aquascaping competitions, was established in 2001 and continues to inspire new generations of aquascapers. These competitions have played a major role in popularizing aquascaping as both a hobby and an art form, with participants exploring various styles, from lush junglescapes to minimalist rock formations.

Today, aquascaping is a global phenomenon, with vibrant communities of enthusiasts who share their designs, techniques, and knowledge online. Social media platforms have become gathering places for aquascapers to showcase their work, exchange tips, and push each other to new creative heights. Whether you're inspired by Amano's nature aquariums or experimenting with your own unique style, aquascaping offers endless possibilities for creative expression.

As you embark on your own aquascaping journey, it's important to remember that you're part of a tradition that stretches back centuries. You're not just setting up an aquarium; you're participating in an art form that blends design, biology, and environmental stewardship. Every aquascape you create adds to the rich history of this evolving discipline, and with each new tank, you help shape the future of aquascaping.

Basic Concepts of Aquascaping: Art Meets Nature

Aquascaping brings together the beauty of art and the wonders of nature, allowing you to design and nurture an aquatic environment that thrives both aesthetically and functionally. In this practice, you become both a gardener and an artist, combining plants, rocks, and water into a seamless, living composition. The beauty of aquascaping lies in its ability to imitate the natural world while allowing for creative expression. As you begin your aquascaping journey, it's essential to understand the basic concepts that will guide you in creating your aquatic masterpiece.

The Rule of Thirds

A key principle borrowed from photography and painting is the rule of thirds, which is a powerful tool in aquascaping. To apply this rule, imagine dividing your aquarium into nine equal parts using two horizontal and two vertical lines, like a tic-tac-toe grid. The idea is to place your focal points, such as a striking piece of driftwood or a vibrant cluster of plants, along these lines or at their intersections.

This simple technique helps balance your design and draw the viewer's eye naturally across the aquascape. For example, if you have a large rock formation, placing it off-center at one of the intersection points can create a dynamic composition. Research in visual psychology has shown that people's eyes naturally gravitate toward these points in an image, which makes the rule of thirds an invaluable tool in creating visually compelling aquascapes.

Hardscape and Softscape: Building the Foundation

In aquascaping, you'll work with two primary elements: hardscape and softscape. Hardscape refers to the non-living components of your design, such as rocks, driftwood, and substrate (the material at the bottom of the tank). These elements form the backbone of your aquascape, providing structure and depth to your layout.

For instance, using a variety of rock sizes can help you create a sense of scale. You might use larger rocks to mimic mountain formations, while smaller stones can suggest the gradual erosion of a riverbed. Driftwood, on the other hand, adds an organic flow to the tank, guiding the viewer's eye and creating natural points of interest.

Softscape includes the living components, primarily plants, mosses, and in some cases, aquatic creatures. Plants can dramatically change the mood and atmosphere of your aquascape. Fast-growing stem plants like Rotala or Cabomba can create a dense, jungle-like look while slow-growing Anubias or Buces give a more minimalist, refined feel. The choice and placement of these plants play a crucial role in defining the overall character of your tank.

The Golden Ratio: Finding Natural Balance

Another concept that can help you achieve a more balanced design is the golden ratio. This mathematical ratio, approximately 1:1.618, has been used in art and architecture for centuries. Found frequently in nature — in the spirals of shells, the branching of trees, and even the proportions of the human face — the golden ratio naturally appeals to the human eye.

In aquascaping, you can apply the golden ratio by dividing your aquarium according to these proportions and placing your key elements at focal points that follow this rule. For instance, positioning a large rock or a vibrant plant cluster at a point that aligns with the golden ratio can create a more harmonious and visually satisfying layout.

Creating Depth and Perspective

One of the biggest challenges in aquascaping is making a small space feel expansive. Since you're working within the confines of an aquarium, you must use clever techniques to create the illusion of depth and perspective. One such technique is layering. Placing larger plants and hardscape materials, such as rocks or driftwood, towards the back of the tank while positioning smaller plants and finer details towards the front helps create a sense of distance.

By doing this, you can draw the viewer's eye deeper into the tank, making it feel as though the aquascape extends far beyond its physical boundaries. Another method involves using a sloped substrate. A gentle slope from the front of the tank to the back can enhance the illusion of depth and make the tank appear more dynamic.

Light and Color: Illuminating Your Design

Lighting is an essential component of aquascaping. It not only supports plant growth but also affects the way your aquascape is perceived. The interplay of light and shadow can add depth and drama to your tank. For example, strategically placing light to highlight a focal point, like a vibrant red plant or an intricate piece of driftwood, can make it stand out more effectively.

In terms of color, aquascapers often use green plants as a backdrop and add pops of contrasting colors, such as red or orange, to create visual interest. Red plants, like Alternanthera Reineckii, can serve as a focal point, drawing the viewer's attention and adding warmth to the overall composition. Just as in painting, color theory plays an important role in guiding the eye and balancing the composition.

Embracing Negative Space

One of the most overlooked aspects of aquascaping is negative space or the empty areas in your tank. It's tempting to fill every inch with plants and hardscape elements, but a cluttered tank can feel chaotic and overwhelming. By intentionally leaving some areas open, you allow the viewer's eye to rest and appreciate the overall composition.

Negative space can also enhance the visual impact of your focal points. A beautifully placed piece of driftwood or a colorful plant stands out even more when contrasted against open areas. Think of negative space as the pauses between notes in a piece of music— it's just as important as the elements themselves in creating a balanced design.

Creating Flow: The Movement of Design

A successful aquascape isn't just a collection of beautiful elements; it's about how those elements interact to create a sense of flow. In nature, nothing is static. Rivers carve through landscapes, plants grow towards the light, and rocks are worn down by the wind. Your aquascape should reflect that natural movement.

One way to create flow is by positioning your hardscape elements in a way that guides the viewer's eye across the tank. For instance, you might arrange rocks and driftwood in a diagonal line to suggest the natural flow of water. The plants, too, should appear to have grown organically, with taller plants in the back and shorter ones in the front, creating a smooth transition.

Bringing it All Together

As you embark on your aquascaping journey, remember that you're not just creating an underwater scene— you're crafting an experience. From the strategic use of negative space to the careful placement of plants and hardscape, every decision you make shapes how your aquascape feels and how others will perceive it. Balancing these elements requires patience, but the result is a living work of art that reflects both the beauty of nature and your personal vision.

The Importance of Research and Planning

While it might be tempting to dive straight into setting up your tank, proper research and planning are crucial to the success of your aquascape. Without a clear plan, you may end up with a tank that doesn't thrive or looks unbalanced. A well-thought-out plan ensures that all the elements in your aquascape will work together harmoniously, both aesthetically and functionally.

Setting Your Vision

Before you start, take some time to think about the overall look and feel you want to achieve. Are you inspired by lush tropical landscapes, or do you prefer a more minimalist, Zen-like aesthetic? Having a clear vision will help guide your decisions on everything from the type of plants and fish you select to the layout of your hardscape.

Choosing the Right Plants and Fish

Before you begin, it's essential to research the plants and fish you plan to include in your tank. Not all species are compatible with each other, and some may require specific water conditions or lighting setups. For instance, certain plants thrive in high-light environments, while others prefer low-light conditions. Similarly, some fish may disturb delicate plants or outcompete other species for food.

You also need to consider the growth rate and maintenance requirements of your chosen plants. Fast-growing plants may need frequent trimming to keep them from overtaking your aquascape, while slow-growing species may require more attention to ensure they stay healthy.

By researching the needs of each species, you can avoid common pitfalls and ensure that your aquascape remains healthy and balanced over time. For example, mosses like Java Moss are ideal for beginners because they're hardy and low-maintenance, while stem plants like Rotala require more frequent pruning and nutrient supplementation.

Lighting and Equipment

Lighting is another critical aspect of aquascaping that often goes overlooked. The type and intensity of light you choose will affect plant growth, algae control, and the overall look of your aquascape. High-light tanks can support more demanding plants but may also encourage algae growth if not properly balanced. Low-light setups are easier to maintain but limit the types of plants you can grow.

In addition to lighting, you'll need to invest in proper filtration and CO_2 systems. A good filtration system helps maintain water quality by removing waste and toxins, while a CO_2 system provides the necessary carbon for plant growth. Without adequate CO_2, your plants may struggle to thrive, leading to poor growth and an imbalance in your ecosystem.

Testing and Adjusting Water Parameters

Water quality plays a significant role in the health of your aquascape. Many beginner aquascapers make the mistake of skipping this step, only to encounter problems down the line with algae blooms or fish health issues. Before introducing any plants or fish, you need to test the water parameters, including pH, hardness, and nutrient levels. Each plant and fish species has its own preferred water conditions, so it's essential to ensure that these parameters are within the acceptable range.

In many cases, you may need to make adjustments to your water before adding life to the tank. For instance, some plants require softer water with a lower pH, while others can tolerate more alkaline conditions. Regular water testing helps you stay on top of any changes that could affect the health of your aquascape.

Designing for Maintenance

Another key aspect of planning is designing your aquascape with maintenance in mind. A beautifully planted tank can quickly become overwhelming if it requires constant trimming and upkeep. Consider how fast your plants will grow, how often they'll need to be pruned, and whether the design allows easy access for cleaning.

Understanding the Ecosystem: Balancing Plants, Fish, and Bacteria

Aquascaping is more than just creating a visually stunning aquarium; it's about fostering a balanced ecosystem where plants, fish, and bacteria coexist in harmony. Each element plays a vital role in maintaining the health of the entire system, and neglecting one can lead to imbalances that affect the whole environment.

The Nitrogen Cycle: A Foundation of Life

At the heart of any aquatic ecosystem is the nitrogen cycle, a process that involves the conversion of ammonia (produced by fish waste and decaying organic matter) into nitrites and then nitrates by beneficial bacteria. Without this cycle, toxic ammonia levels would build up, harming or killing your fish and plants.

Beneficial bacteria are crucial to this process, breaking down harmful substances and converting them into less toxic forms that can be absorbed by plants. Ensuring that your tank has a healthy population of these bacteria is essential for maintaining water quality and preventing dangerous ammonia spikes.

You can establish this bacteria colony by cycling your tank before introducing fish. This involves adding a source of ammonia, such as fish food or liquid ammonia, and allowing the bacteria to grow over a few weeks. Regular testing will let you know when your tank is fully cycled and ready for fish.

Plants as the Powerhouses

In an aquascape, plants are more than just decorative elements — they are the powerhouses that drive the ecosystem. Through photosynthesis, plants produce oxygen, which is vital for the survival of fish and other aquatic organisms. They also absorb excess nutrients in the water, helping to prevent algae growth and keeping the water clean.

But not all plants are created equal. Some, like fast-growing stem plants, are excellent at absorbing nutrients and preventing algae, while others, like mosses, provide hiding places for fish and shrimp but require more frequent maintenance.

The Role of Fish

Fish are more than just the stars of the show; they contribute to the ecosystem in several ways. Fish waste provides essential nutrients for plants, and their natural behaviors can help maintain balance in the tank. For example, bottom-dwelling fish like Corydoras stir up the substrate, preventing debris from accumulating and encouraging nutrient cycling.

However, it's crucial to avoid overstocking your tank. Too many fish can lead to excessive waste production, which can overwhelm the plants and bacteria and cause water quality issues. A good rule of thumb is to start with a few fish and gradually add more as your tank stabilizes.

Achieving Balance

Balancing plants, fish, and bacteria is a delicate process, and it requires ongoing monitoring and adjustments. If you notice algae blooms, it could be a sign that your plants aren't absorbing enough nutrients or that your light levels are too high. Similarly, if your fish seems stressed or lethargic, it may indicate poor water quality or imbalanced water parameters.

Furthermore, experts recommend starting with a small number of fish and gradually increasing the population as your plants establish themselves. This allows the ecosystem to adjust and helps prevent any sudden changes that could disrupt the balance.

Regular testing, maintenance, and observation are key to maintaining a thriving aquascape. Over time, you'll develop an intuitive sense of how your tank operates, allowing you to make small adjustments that keep everything in harmony.

The Bottom Line

Aquascaping is a journey of creativity, patience, and learning. As you venture into the world of underwater design, you'll find that every choice — from the placement of a rock to the type of fish you add — contributes to the overall health and beauty of your aquascape. It's an art form that challenges you to blend nature and design while fostering life in a miniature ecosystem.

Remember, the best aquascapes don't happen overnight. They evolve over time, with careful planning, research, and a commitment to nurturing your underwater garden. By understanding the history, mastering the basic concepts, and balancing the delicate relationships between plants, fish, and bacteria, you'll be well on your way to creating a stunning and thriving aquascape that you can proudly call your own.

CHAPTER 02

TYPES OF **FRESHWATER** AQUARIUMS

"Freshwater aquariums can range from simple community tanks to complex biotope aquariums, each type offering its unique challenges and rewards."

Christel Kasselmann

Stepping into the world of aquascaping opens up a realm of possibilities, but with so many choices, it can be overwhelming to decide where to begin. What type of aquarium is best for your space, what is your experience level, and what amount of time are you willing to invest in maintaining it? Whether you're drawn to lush, planted tanks, realistic recreations of natural habitats, or intricate nano aquariums, understanding the different types of freshwater aquariums can help you find the setup that fits your vision and lifestyle.

This chapter is designed to guide you through the various types of freshwater aquariums, each with its unique features, benefits, and challenges. We'll dive deep into the differences between low-tech and high-tech aquariums, explore the elegance of planted tanks, discuss biotope aquariums for nature lovers, and introduce the captivating world of nano tanks. We'll also look at hybrid styles like paludariums and ripariums that combine elements of both land and water.

By the end of this chapter, you'll have a solid understanding of which aquarium style suits your needs and preferences, allowing you to start your aquascaping journey with confidence.

Low-Tech vs. High-Tech Aquariums

When you're just starting out with aquascaping, one of the key decisions you'll face is whether to opt for a low-tech or high-tech aquarium. This choice shapes everything from the types of plants you can grow to the level of maintenance your tank will require. Both options offer distinct advantages and challenges, so understanding the differences will help you select the setup that aligns with your goals and lifestyle.

Low-Tech Aquariums: Simplicity and Balance

Low-tech aquariums are often seen as the go-to option for beginners, but don't be fooled into thinking they're only for newcomers. These setups can be perfect for anyone looking to maintain a beautiful aquarium without the need for advanced technology. The main feature of a low-tech aquarium is its minimal reliance on equipment. You typically avoid using CO2 injection systems and powerful lighting. Instead, low-tech tanks rely on hardy plant species that thrive in lower light conditions without the need for added CO2.

Pros of Low-Tech Aquariums

The following are noteworthy pros of using low-tech aquariums:

- **Lower Cost:** One of the main draws of a low-tech aquarium is its affordability. Without the need for high-end lighting systems, CO2 injection, or complex filtration systems, the initial cost is much lower. This makes it an attractive option if you're just getting started and don't want to invest heavily upfront.
- **Less Maintenance:** Low-tech tanks are often easier to maintain. They typically require less frequent water changes and plant trimming, which is great if you have a busy schedule. Plants grow more slowly, reducing the need for constant attention, which allows your aquarium to maintain a natural, balanced rhythm.
- **Simplicity:** If you're looking for a straightforward experience, low-tech aquariums are more forgiving. They're great for hobbyists who want a low-maintenance setup that doesn't demand daily care or adjustments.

- **Natural Growth:** Without the artificial boost of CO_2 and high-powered lights, plants grow at a slower, more natural pace. This creates a more organic look over time, allowing the aquascape to mature gradually.

Cons of Low-Tech Aquariums

However, like anything, low-tech aquariums have their challenges, too. Here are a few downsides to using low-tech aquariums:

- **Limited Plant Selection:** Because you're working with lower light and no CO_2, your choices in plant species are limited. While there are many beautiful hardy plants like Anubias and Java Fern, you may find the options less varied compared to a high-tech setup.
- **Slower Growth:** If you're hoping for fast results, a low-tech tank might feel a bit sluggish. Plants grow at a slower rate, which can be frustrating if you're eager to see your aquascape take shape quickly.
- **Potential for Algae:** Since plant growth is slower, there's less competition for nutrients in the water. If you're not careful with your light levels and nutrient balance, algae can easily take over. Managing this requires careful observation and adjustments to avoid an unsightly bloom.

High-Tech Aquariums: A Faster-Paced, Dynamic Experience

On the other hand, high-tech aquariums are the choice for hobbyists who want to push the boundaries of what's possible in aquascaping. These tanks use advanced technology like powerful LED lighting, CO_2 injection systems, and external filters to support fast-growing plants and create vibrant, dynamic aquascapes. While they require more time and effort, they also offer greater creative flexibility.

Pros of High-Tech Aquariums

Here are a few reasons why you may want to try high-tech aquariums:

- **Broader Plant Selection:** With the addition of CO_2 and strong lighting, the door opens to a wide variety of plant species, including more delicate and vibrant types like Rotala or Hemianthus callitrichoides (commonly known as dwarf baby tears). This gives you more flexibility in your designs and allows for more intricate, lush aquascapes.
- **Faster Growth:** The combination of CO_2, high-intensity lighting, and nutrient dosing means that plants grow quickly. You'll see changes in your aquascape within weeks, which can be exciting if you enjoy

tweaking and experimenting with different designs.

- **Greater Design Flexibility:** High-tech tanks allow for more elaborate, densely planted landscapes. You can create intricate layouts that would be difficult, if not impossible, in a low-tech setup.

Cons of High-Tech Aquariums
Despite the foregoing benefits, high-tech aquariums have some downsides, which include the following:

- **Higher Cost:** The downside to all this technology is the cost. High-tech tanks require a significant investment in equipment, from CO2 systems to advanced lighting and filtration. If you're on a tight budget, the upfront cost of a high-tech aquarium might be a barrier.
- **Increased Maintenance:** With faster plant growth comes more frequent pruning, water changes, and monitoring of water parameters. A high-tech setup requires constant attention to ensure everything is balanced. You might find yourself dedicating several hours each week to maintaining the tank.
- **Steeper Learning Curve:** If you're new to aquascaping, the complexity of balancing CO2 levels, light intensity, and nutrients can feel overwhelming. High-tech setups require careful attention to detail, as an imbalance can lead to issues like algae outbreaks or plant die-offs. If you're not prepared to spend time learning the ins and outs, it can quickly become frustrating.

Which One is Right for You?
Choosing between a low-tech and high-tech aquarium depends on your experience level, budget, and how much time you're willing to invest in maintenance. If you're new to aquascaping or prefer a more relaxed, hands-off approach, a low-tech setup is a great way to get started. The simplicity and lower cost make it ideal for beginners or those with busy schedules.

On the other hand, if you're excited by the idea of creating a fast-growing, vibrant aquascape and don't mind dedicating time to maintenance, a high-tech tank offers greater creative possibilities. You'll have access to a broader range of plant species and the ability to design intricate, lush landscapes. However, be prepared for the investment—both in terms of money and time.

Ultimately, there's no right or wrong choice—just the one that fits your needs and goals. Some hobbyists start with a low-tech tank to gain experience, then transition to high-tech setups as they become more comfortable. Others stick with low-tech aquariums for years, appreciating the simplicity and natural beauty they offer.

Whichever path you choose, both options allow you to immerse yourself in the rewarding world of aquascaping, where art and nature come together to create something truly unique.

Planted Aquariums

Planted aquariums are among the most captivating and rewarding styles in the aquascaping world. When you create a planted tank, you're not just designing a space for fish—you're crafting a vibrant, dynamic ecosystem where plants take center stage. These tanks go beyond being simple aquatic environments; they become living, breathing landscapes that echo the beauty and complexity of the natural world. Whether you're recreating a lush jungle, a peaceful riverbank, or a well-tended garden, the planted aquarium offers endless possibilities for creative expression.

Dutch Style: A Garden Underwater

One of the oldest and most structured styles of planted aquariums is the Dutch style, which originated in the Netherlands. This style is perfect if you love precision, order, and plant variety. A Dutch-style tank focuses on dense, vibrant plant growth arranged in layers and blocks that resemble a carefully planned garden. The emphasis here is not on hardscape elements like rocks or driftwood but rather on the diversity and arrangement of plants.

In a Dutch-style aquarium, you'll often see plants grouped into what is known as "street" planting, where rows of plants are organized to create depth and perspective. The use of contrasting colors and textures is key to making the plants stand out. For example, you might see bright red Ludwigia next to a dark green Cryptocoryne, creating a visual pop that draws the eye across the tank.

This style is one of the most maintenance-heavy, requiring frequent pruning and careful attention to plant health. However, the reward is a stunning underwater garden that is full of life and color. In a way, it's like having a living piece of art in your home, constantly evolving as the plants grow and change.

Nature Style: Mimicking Natural Landscapes

If you prefer a more organic and free-flowing look, the Nature style might be your ideal choice. This style was popularized by Japanese aquascaper Takashi Amano, whose philosophy was to mimic the randomness and beauty of nature. In a nature-style aquarium, you're encouraged to let your creativity flow by arranging rocks, driftwood, and plants in ways that resemble natural environments like forests, rivers, or mountains.

One of the signature techniques of the nature style is the use of the golden ratio or the rule of thirds to create balanced, visually appealing compositions. For instance, you might place a large rock or piece of driftwood off-center, following these proportions, to create a focal point that feels harmonious and natural.

Amano often took inspiration from Japanese Zen gardens, and many nature-style tanks evoke a sense of calm and tranquility. Plants in this style are allowed to grow more freely, without strict pruning, creating a landscape that feels untouched by human hands. This style is perfect if you want your aquarium to reflect the serene, chaotic beauty of the natural world, and it provides a great deal of creative freedom.

Iwagumi: Simplicity and Precision

The Iwagumi style is one of the most minimalist approaches to aquascaping, but don't let its simplicity fool you—it can be one of the most challenging to master. Originating in Japan, "Iwagumi" translates to "rock formation," and the primary focus of this style is the careful placement of stones. In an Iwagumi aquarium, the rocks are the stars of the show, while plants play a secondary role in complementing the hardscape.

Typically, an Iwagumi tank will feature three main stones: a dominant stone, a secondary stone, and a smaller supporting stone. These rocks are arranged according to principles of Japanese design, aiming to create balance and harmony. The idea is to evoke a natural, peaceful landscape, often resembling mountain ranges or riverbeds.

Plants in an Iwagumi tank are usually limited to one or two species, with Carpet plants like Hemianthus Callitrichoides (dwarf baby tears) often used to cover the substrate. The minimalistic approach may seem easy, but achieving the perfect layout requires precision and a keen eye for detail. However, when done right, an Iwagumi aquascape can be breathtaking in its simplicity, drawing the viewer's attention to the subtle beauty of rock and plant interplay.

Jungle Style: Embracing Wild Beauty

At the opposite end of the spectrum from Iwagumi is the Jungle style, a much more relaxed and freeform approach. As the name suggests, jungle-style aquariums aim to recreate the dense, overgrown look of a tropical jungle. This style embraces the wild, untamed side of nature, allowing plants to grow freely and in abundance.

A jungle-style aquascape is typically filled with tall, fast-growing plants that create a lush canopy effect. You'll often find a mix of species like Amazon Sword, Java Fern, and creeping plants that cover the substrate, giving the tank a dense, layered look. The goal is to make the aquarium feel alive and vibrant, as though it could exist naturally in the wild.

While this style may appear effortless, it still requires careful planning. Without proper attention, the fast-growing plants can easily overwhelm the tank, blocking out light and creating dead zones where other plants struggle to survive. However, the jungle style offers a lot of creative freedom, as it doesn't adhere to strict rules or patterns. It's ideal if you want your aquarium to feel immersive and full of life.

Crafting Your Own Underwater Paradise

Choosing a planted aquarium style comes down to your personal preferences and the level of care you're willing to invest. Whether you're drawn to the structured beauty of a Dutch tank, the organic flow of a nature-style setup, the minimalist elegance of Iwagumi, or the wild freedom of a jungle tank, there's no shortage of inspiration.

Planted aquariums are more than just a hobby—they're a way to bring a piece of the natural world into your home. As you nurture your aquatic plants, you'll find that your tank becomes a living, evolving work of art that reflects both the beauty of nature and your personal creativity.

Biotope Aquariums

A biotope aquarium is a unique and fascinating way to recreate nature in your home. Unlike other styles of aquascaping that focus on aesthetics, biotope aquariums aim to replicate a specific natural habitat as accurately as possible. This means that every element of the tank—from the fish and plants to the water conditions and substrate—must match the real-world ecosystem you're mimicking. It's an exciting challenge that goes beyond simply creating a beautiful aquascape. Here, you're crafting an authentic, living snapshot of a natural environment.

Recreating Natural Habitats for Fish and Plants

When you choose to set up a biotope aquarium, you're taking on the task of recreating the exact conditions of a specific location where certain fish and plants thrive in the wild. This requires detailed research and careful planning, as the goal is not just visual beauty but functional accuracy. You'll need to match the water parameters, such as pH, hardness, and temperature, to those found in the original habitat. The types of plants, fish, and even the substrate must all be appropriate for the biotope you are recreating.

For example, if you're drawn to the Amazon River biotope, you'll need soft, acidic water, which is characteristic of that region. The substrate might include a mix of fine sand and leaf litter to simulate the natural riverbed, while driftwood adds a natural touch, reflecting the fallen branches and tree roots typically found in Amazon waterways. Plants such as Amazon swords and floating species like Pistia can be used to recreate the lush greenery of the Amazon basin. The fish you select should also be native to that region, with species like tetras, angelfish, and corydoras thriving in this setup.

The Challenge and Reward

Creating a biotope aquarium can be more challenging than other aquascaping styles because it requires a deep understanding of the habitat you're replicating. You need to consider not only the aesthetics but also the specific needs of the fish and plants. For example, recreating the water chemistry of the Amazon or maintaining the correct lighting for a Southeast Asian stream can be difficult, but it's essential for keeping your aquarium inhabitants healthy.

However, the challenge is part of what makes biotope aquariums so rewarding. There's something special about knowing you've successfully recreated a miniature version of a real-world ecosystem. Watching your fish and plants thrive in an environment that mirrors their natural habitat can be incredibly satisfying. It's more than just an aquarium—it's a window into the wild.

To sum it up, a biotope aquarium is a beautiful and rewarding way to experience nature in your home. By carefully researching and replicating a specific habitat, you can create a thriving ecosystem that's not only visually stunning but also true to the natural world.

Nano Aquariums

Nano aquariums, typically defined as tanks under 30 liters (7 gallons), have become increasingly popular among aquascaping enthusiasts in recent years. These small-scale tanks are perfect for hobbyists who want to enjoy the beauty of aquascaping without taking up too much space. While their compact size makes them appealing, nano tanks present unique challenges that both beginners and experienced aquascapers need to consider. This subtopic fits within the broader discussion of freshwater aquariums by exploring a specialized niche—one that blends practicality with artistic potential.

Small-Scale Aquascaping for Beginners and Advanced Hobbyists

At first glance, nano aquariums seem like an ideal option for those new to aquascaping. After all, a smaller tank means less water, fewer materials, and a reduced need for large-scale equipment. Nano tanks are especially appealing if you're working with limited space, whether it's a small apartment, office desk, or even a bookshelf. They allow you to experiment with different styles without the commitment of maintaining a larger tank. However, while the size may be smaller, the complexities of keeping a nano aquarium healthy can be greater.

The limited water volume in a nano tank means that even small changes in water parameters—such as pH, temperature, or nutrient levels—can have a significant impact. For example, in a 100-liter tank, a small temperature fluctuation might go unnoticed, but in a nano aquarium, even a minor shift can be dangerous for fish and plants. This means you have to be extra vigilant about maintaining consistent conditions. A spike in ammonia or a drop in pH can occur quickly in such a small environment, requiring prompt action to prevent damage to the tank's ecosystem.

Key Considerations for Nano Aquariums

While nano tanks are visually impressive, they require a high level of care and attention. The most important factor is consistency. Because of the small water volume, weekly maintenance becomes crucial. You'll need to regularly monitor water parameters, conduct partial water changes, and carefully manage feeding to avoid overloading the system with waste. Additionally, temperature control is critical. Even a small shift in room temperature can cause fluctuations that might stress or harm the tank's inhabitants.

It's also important to consider the size and species of the fish you introduce. Nano tanks are ideal for smaller species like betta fish, neon tetras, or pygmy corydoras, but overcrowding can lead to poor water quality and stressed fish. The limited space makes every decision more significant, requiring careful planning and attention to detail.

Paludariums, Ripariums, and Emergent Aquascapes

When it comes to freshwater aquariums, many hobbyists focus on creating lush underwater landscapes. However, if you're looking to expand beyond the traditional aquascape, hybrid setups like paludariums, ripariums, and emergent aquascapes offer exciting new possibilities. These styles combine aquatic and terrestrial elements, creating dynamic, multi-layered environments that are not only beautiful but also unique. By blending land and water, these setups mimic natural habitats more closely, offering diverse ecosystems within a single tank.

Paludariums

A paludarium is a hybrid aquarium that integrates both aquatic and terrestrial elements. In essence, it's like having an aquarium and a terrarium in one setup. This style is ideal for hobbyists who want to create a dynamic display featuring fish, amphibians, reptiles, and a wide variety of plants. The tank is often divided into a submerged area for fish and aquatic plants and an elevated, dry section for terrestrial plants, amphibians like frogs or newts, and even small reptiles like geckos.

Creating a paludarium requires careful planning. You need to consider how to balance both the water and land environments. For example, humidity levels must be high enough to support moisture-loving plants like mosses and ferns while also ensuring the water area is clean and well-filtered for aquatic species. Lighting must also accommodate both the underwater and terrestrial areas, which can be a challenge. Special LED lights that provide full-spectrum coverage often work well in these setups.

What's fascinating about paludariums is the variety of species you can incorporate. For example, crabs like fiddler crabs can thrive in the water while using the land area to bask. These tanks offer an immersive experience that allows you to observe interactions between water and land creatures, creating a miniature world that feels alive and constantly changing.

Ripariums

A riparium focuses on recreating the look and feel of riverbanks or wetlands, emphasizing semi-aquatic plants that grow both above and below the water. In a riparium, plants like Anubias or Java ferns have their roots submerged in water, but their leaves rise above the surface. This setup allows you to mimic the lush greenery you'd see along the edges of rivers, ponds, or wetlands.

Ripariums offer a unique visual experience that combines the structure of an underwater landscape with the natural beauty of an emergent plant environment. For instance, Pothos or Spathiphyllum can be grown with their roots submerged, creating a striking, overgrown effect as their leaves spread above the tank. The key to a successful riparium is maintaining the delicate balance of moisture levels, ensuring that the plants thrive both underwater and above it.

These tanks tend to evoke a calm, natural aesthetic, making them perfect for those who want a tranquil and visually interesting addition to their home. Fish species such as gouramis or small tetras often do well in ripariums, complementing the lush plant life above them.

Emergent Aquascapes

Emergent aquascaping takes the concept of a traditional planted aquarium and pushes it further by encouraging plants to grow above the water's surface. This style is often seen in nature-inspired aquascapes, where plants like mosses, bucephalandra, and ferns are allowed to extend beyond the waterline, clinging to rocks, driftwood, or other hardscape elements.

Emergent plants can transform the look of your aquascape, creating the illusion of waterfalls or rocky outcrops seen in the wild. For example, placing driftwood that rises above the surface allows plants to climb upward, blurring the line between land and water. The visual effect can be dramatic, especially when the emergent growth is paired with mosses that cascade down from the rocks into the water below.

The key to success in emergent aquascaping is ensuring both the submerged and emergent plants receive enough light and moisture. Some plants may require higher humidity levels above the water, so it's important to adjust the tank's environment accordingly.

The Beauty of Hybrid Aquascapes

Paludariums, ripariums, and emergent aquascapes offer exciting opportunities to break away from traditional underwater-only setups. Whether you want to create a mixed environment for both aquatic and terrestrial species in a paludarium, mimic a riverbank in a riparium, or blur the lines between land and water with emergent aquascaping, these hybrid styles allow for endless creativity. With the right balance of light, moisture, and plant care, your tank can become a thriving, dynamic ecosystem that showcases the beauty of both land and water.

The Bottom Line

Freshwater aquascaping is a rich and diverse hobby, with countless ways to design and maintain your tank. Whether you're drawn to the simplicity of a low-tech setup, the complexity of a biotope aquarium, or the dynamic interplay of land and water in a paludarium, there's a style out there that's perfect for you. As you explore these different types of aquariums, remember that the most important thing is to create a setup that reflects your vision and provides a healthy, thriving environment for your plants and fish. Happy aquascaping!

CHAPTER 03

ESSENTIAL **EQUIPMENT** AND TOOLS FOR **AQUASCAPING**

"The right tools make a huge difference in aquascaping."

Karen Randall

When you're just starting your journey into aquascaping, one of the first things you'll realize is that it's more than just arranging plants and fish in a tank. To create a thriving, beautiful underwater ecosystem, you need the right equipment and tools. From selecting the perfect tank to choosing lighting systems and filtration setups, every decision you make will directly impact the success of your aquascape. But don't worry — this chapter will guide you through each piece of essential equipment you'll need, breaking it down in a way that's easy to understand and apply to your own aquarium.

Aquarium Tank Selection

Your aquarium tank is the foundation of your aquascape. It's where your vision comes to life, and choosing the right one can either simplify or complicate your entire project. When deciding on a tank, you'll need to consider its size, shape, and the material it's made from. Each of these factors can influence how your aquascape looks and functions.

Size

The size of your aquarium will significantly impact your aquascape's complexity and maintenance requirements. Larger tanks generally provide more flexibility in terms of design and are more forgiving when it comes to water quality, as the larger volume of water can help buffer against sudden changes in parameters. Tanks over 20 gallons are often recommended for beginners, as they offer ample space for plants, fish, and hardscape materials without becoming overwhelming to manage. Smaller tanks, such as nano aquariums (under 10 gallons), present unique challenges. While they require less space and can be a beautiful addition to small homes or offices, their smaller water volume makes them more susceptible to fluctuations in temperature and water chemistry. If you're a beginner, starting with a mid-sized tank, like a 20- to 40-gallon aquarium, offers a balanced introduction to aquascaping without being too demanding.

Shape

Aquarium tanks come in a variety of shapes, from standard rectangles to more unconventional designs like cubes or bow-front tanks. Rectangular tanks are the most popular and versatile because they offer a large surface area for planting and easy access for maintenance. Cubes, on the other hand, offer a compact design that's perfect for smaller spaces, though they can sometimes limit your options for creating depth and perspective in your aquascape.

If you're looking for something more visually dynamic, a bow-front tank provides a curved front glass panel, giving you a panoramic view of your aquascape. However, this can distort the view slightly, making it harder to appreciate the finer details of your design.

Materials

Aquarium tanks are typically made from two materials: glass or acrylic. Glass tanks are more common because they are durable, scratch-resistant, and provide excellent clarity. They're also less likely to yellow over time, ensuring your aquascape remains clear and vibrant. However, glass tanks tend to be heavier and more prone to breaking.

Acrylic tanks, on the other hand, are lighter and more impact-resistant, which makes them a good choice for larger tanks or when weight is a concern. The downside is that acrylic scratches more easily, and over time, these scratches can dull the tank's appearance. The choice between glass and acrylic ultimately comes down to your priorities: durability and weight versus clarity and scratch resistance. When choosing the material, consider your budget, the size of the tank, and whether the tank's weight or shape will be an issue. For most hobbyists, a standard glass tank offers the best balance of cost, clarity, and durability.

Lighting Systems

Once you've selected your tank, the next crucial step is to provide the right lighting. Lighting is one of the most critical aspects of a successful aquascape, especially if you're growing live plants. The type, intensity, and duration of light all influence plant growth, coloration, and overall health. Selecting the right lighting system can feel daunting, but with a solid understanding of your options, you'll be able to make an informed choice that supports both the aesthetic and biological needs of your aquascape.

LED vs. Fluorescent

When it comes to aquarium lighting, the two most common options are LED and fluorescent systems. Let's briefly review the difference between the two types of lighting:

LED Lighting

LED (light-emitting diode) lighting has become the gold standard in modern aquascaping. It's energy-efficient, produces less h eat, and offers a wide range of customization options. One of the most significant advantages of LED lights is the ability to adjust the color spectrum, which can help plants grow while also enhancing the visual aesthetics of your tank.

LED fixtures are available in a variety of intensities and sizes, making them suitable for everything from small nano tanks to large, high-tech aquariums. Many LED systems also come with programmable settings that allow you to simulate natural lighting conditions, such as sunrise, sunset, and even moonlight, which can create a more dynamic and natural environment for your plants and fish.

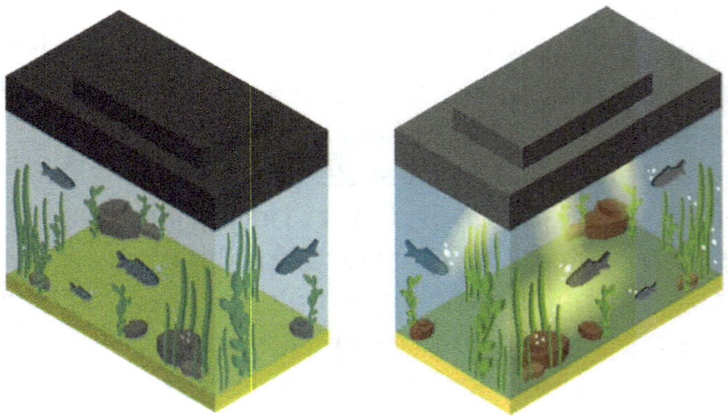

Fluorescent Lighting

While LED lights dominate the market, fluorescent lights, particularly T5 and T8 fluorescent tubes, are still a viable option for aquascaping. They are generally less expensive than LED lights and can provide adequate lighting for low- to medium-light plants. However, fluorescent lighting is less energy-efficient and has a shorter lifespan compared to LEDs.

Fluorescent lights also produce more heat, which may affect the temperature of your aquarium, especially in smaller tanks. Additionally, they are not as customizable as LED systems in terms of light spectrum and intensity, which can limit their usefulness for advanced aquascaping setups.

Choosing the Right Light for Your Plants

In summary, LED lighting is typically the best option for most aquascapers due to its efficiency, longevity, and adaptability. However, if you're on a tight budget, fluorescent lighting can still get the job done, especially in lower-tech setups.

Understanding Light Spectrum, Intensity, and Duration

The health and growth of plants in an aquascape depend heavily on the quality, strength, and timing of the light they receive. Light spectrum, intensity, and duration are key factors that work together to create an ideal environment for photosynthesis. By mastering these elements, you can ensure that your plants grow robustly, with vibrant colors and healthy foliage.

- **Light Spectrum:** Plants rely on specific wavelengths of light for photosynthesis. The best lighting systems for aquascaping offer a full spectrum of light that includes both blue and red wavelengths, which are crucial for plant growth. Blue light encourages photosynthesis and is particularly important for promoting healthy foliage. Red light, on the other hand, supports flowering and overall plant vitality. A well-balanced spectrum ensures your plants have everything they need to thrive.
- **Light Intensity:** The intensity of light refers to how much light is available for your plants, and this can vary depending on the type of plants you're growing. Low-light plants like Anubias and Java fern require less intensity, while high-light plants like carpeting species (such as dwarf hairgrass) need stronger lighting to grow properly. Matching the light intensity to your plants' needs is crucial to avoid problems like stunted growth or excessive algae.
- **Light Duration:** How long you leave your lights on each day is just as important as the type and intensity of light. Most aquascapers aim for a photoperiod of 8–10 hours per day. Too much light can lead to algae problems, while too little can stunt plant growth. It's also important to provide a consistent light schedule, as plants thrive on regular day and night cycles.

Whether you're working with low-light plants or high-light species, finding the right balance in lighting will help you maintain a thriving aquascape without running into common issues like stunted growth or algae overgrowth.

Filtration Systems

A good filtration system is essential for maintaining water quality and ensuring a healthy environment for your plants and fish. Filters help remove physical debris, harmful chemicals, and excess nutrients, all while promoting the growth of beneficial bacteria. There are several types of filtration systems to choose from, each with its own set of advantages and disadvantages. Let's briefly review them.

Internal Filters

Internal filters sit inside the aquarium and are typically compact, making them a good option for smaller tanks or nano aquariums. They are easy to install and maintain and effective at providing mechanical and biological filtration. However, they can take up valuable space within the tank, which may interfere with the aesthetics of your aquascape and may not be powerful enough for larger or heavily stocked setups.

External Filters

External filters, such as hang-on-back (HOB) filters, are mounted outside the tank, and tubes are used to draw water in, filter it, and return it to the tank. Ideal for larger aquariums, these filters are generally more powerful than internal filters and can handle larger volumes of water. They are easy to maintain and provide excellent mechanical and biological filtration without taking up space inside the tank. Since they don't take up space inside the tank, they allow for more creative aquascaping designs and are a good option for heavily stocked tanks or high-tech aquascapes. However, they can be more expensive and require more effort to set up.

Sponge Filters

Sponge filters are a great option for small tanks, fry tanks, or shrimp tanks, as they provide gentle filtration without creating strong currents. These filters use an air pump to pull water through a sponge, which captures debris and provides a surface for beneficial bacteria to grow. While they are not the most powerful option, sponge filters are cheap, easy to maintain, and ideal for delicate species. However, while sponge filters are an excellent choice for low-tech, low-maintenance tanks, they may not provide enough filtration for heavily stocked or high-tech setups.

Canister Filters

Canister filters are the powerhouse of filtration systems. They are external filters that offer high water flow rates and can accommodate multiple types of filter media, including mechanical, chemical, and biological filtration. Canister filters are ideal for larger aquariums and heavily stocked tanks because they provide excellent water quality. The downside is that they are more expensive and require more maintenance than simpler filters.

Proper Placement and Maintenance

Regardless of the type of filter you choose, proper placement is crucial to ensure effective water circulation. Poor circulation can lead to dead zones where debris and waste accumulate, leading to poor water quality and algae growth. Ensure that your filter's intake and outflow are positioned to promote good water movement, and consider adding additional powerheads or circulation pumps in larger tanks. For internal and sponge filters, placing them near the middle of the tank or where the flow can reach all areas is essential. External and canister filters should be set up so that water flows smoothly throughout the tank without creating dead spots.

Maintenance is another key factor. Regularly cleaning your filter media, whether it's sponges, bio-balls, or activated carbon, ensures that your filter continues to perform effectively. A clogged filter can lead to poor water quality, which in turn can harm your plants and fish.

CO_2 Systems

Carbon dioxide (CO_2) is a vital element for plant growth, and in aquascaping, CO_2 supplementation can make a dramatic difference in the health and appearance of your plants. While some low-tech setups don't require CO_2 injection, adding CO_2 can significantly boost plant growth, especially in high-tech tanks with strong lighting.

CO_2 Injection: Do You Need It?

CO_2 injection can be a game-changer, but it's not always necessary. If you're working with low-light, slow-growing plants, they may get all the CO_2 they need from the natural respiration of fish and other organisms in the tank. However, if you're aiming for a lush, fast-growing aquascape, especially with carpeting plants, CO_2 injection can dramatically improve plant health and growth.

Pros

- **Faster, healthier plant growth:** CO_2 injection accelerates the growth rate of plants, allowing them to absorb nutrients more effectively.
- **Increased plant variety:** CO_2 injection enables you to grow a wider range of plant species, including those that are more demanding.
- **Improved algae control:** Healthy, fast-growing plants outcompete algae for nutrients, helping to keep algae under control.
- **Vibrant coloration:** Many plants develop more intense colors when given enough CO_2.
- **Increased oxygen production:** Healthier plants produce more oxygen, which benefits fish and other aquatic life.

Cons

- **Cost:** CO_2 systems can be expensive, especially for larger tanks, as you'll need a pressurized CO_2 tank, a regulator, and a diffuser.
- **Complexity:** CO_2 injection requires careful monitoring and adjustment to maintain the right balance. Too much CO_2 can harm fish, while too little may not benefit the plants.
- **Maintenance:** CO_2 systems require regular refills and calibration, adding to the overall maintenance of your aquarium.
- **Potential for algae:** If not properly balanced with light and nutrients, excess CO_2 can contribute to algae growth.

DIY vs. Commercial CO_2 Systems

For hobbyists on a budget, DIY CO_2 systems offer a cheaper alternative to commercial setups. These systems typically involve mixing yeast, sugar, and water in a bottle to produce CO_2, which is then diffused into the tank. While effective in smaller tanks, DIY systems can be inconsistent, as CO_2 production varies based on temperature and the fermentation process.

Commercial CO_2 systems, on the other hand, are more reliable and easier to control. These setups usually involve a pressurized CO_2 tank, a regulator to control the flow of CO_2, and a diffuser to disperse the gas into the tank. Although they're more expensive upfront, commercial systems provide consistent CO2 levels and are ideal for larger tanks or high-tech setups.

Substrates and Hardscape Materials

Choosing the right substrate and hardscape materials is critical to creating a stable foundation for your aquascape. The substrate not only anchors your plants but also provides essential nutrients, while hardscape elements like rocks and wood help define the visual structure of your aquarium.

Choosing the Right Substrate for Your Plants

The substrate you choose will depend on the types of plants you want to grow and the overall look you're going for.

- **Inert Substrates:** These include gravel and sand, which don't provide nutrients to plants. Inert substrates are ideal for hardscapes where plants aren't the main focus or when you plan to use root tabs or liquid fertilizers to provide nutrients.
- **Nutrient-Rich Substrates:** These are specially designed for planted aquariums and contain essential nutrients like nitrogen, potassium, and phosphorus. Common brands include ADA Aqua Soil, Fluval Stratum, and Eco-Complete. These substrates support vigorous plant growth and can reduce the need for liquid fertilizers, especially in the early stages of your aquascape.

Rocks, Wood, and Decorations: Natural vs. Synthetic Materials

Hardscape elements like rocks and wood play a vital role in shaping your aquascape. You can choose from natural materials, such as Seiryu stone and driftwood, or synthetic alternatives that mimic the appearance of natural elements.

- **Natural Materials:** Real rocks and driftwood offer authenticity and contribute to the overall balance of your ecosystem. They can also alter water parameters slightly, such as lowering pH in the case of certain types of wood. When using natural materials, it's important to clean and soak them thoroughly to remove any debris or harmful chemicals.
- **Synthetic Materials:** Artificial rocks and wood are typically made from resin and are designed to mimic the appearance of natural elements. They don't affect water chemistry and can be easier to manage, but they may lack the organic beauty of real materials.

Additional Tools and Accessories

Beyond the core equipment like tanks, lights, and filters, there are several other tools and accessories that will make your aquascaping experience more successful and enjoyable. Here are a few essential tools you should consider getting as you embark on your aquascaping journey.

- **Thermometers and Heaters:** Maintaining a stable water temperature is crucial for both plants and fish. A good-quality heater with a built-in thermostat ensures that your tank stays within the optimal temperature range. A thermometer, whether digital or traditional, helps you monitor this.
- **Test Kits:** Regularly testing your water for pH, ammonia, nitrites, and nitrates is essential for maintaining a healthy aquarium. Liquid test kits are often more accurate than test strips, giving you a clearer picture of your water quality.
- **Fertilizers:** In addition to CO2, your plants will need macro- and micronutrients to grow. Liquid fertilizers like Seachem Flourish provide these nutrients, while root tabs can be placed directly on the substrate to feed root-hungry plants.
- **Water Conditioners:** Tap water often contains chlorine and heavy metals that can be harmful to fish and plants. Water conditioners like Seachem Prime neutralize these chemicals, making your water safe for aquarium use.

- **Scissors:** Aquascaping scissors come in different shapes, including straight, curved, and wave. These tools allow you to trim plants easily and precisely, keeping your tank looking neat.
- **Tweezers:** Tweezers are invaluable for planting small or delicate plants, particularly in hard-to-reach areas. They give you greater control and help prevent damage to plant roots.
- **Siphons:** A siphon is used to change water and clean the substrate. Regular water changes are essential for removing excess nutrients and waste that can lead to algae growth.

The Bottom Line

Aquascaping is as much about the right tools and equipment as it is about creativity and design. By understanding the essential components that go into setting up and maintaining a thriving aquarium, you're setting yourself up for success. From selecting the perfect tank to choosing the right lighting and filtration systems, each piece of equipment plays a vital role in the health and beauty of your aquascape.

As you embark on this journey, remember that creating an aquascape is a balance between art and science. Take the time to research, plan, and select the right tools for your specific needs. With the proper equipment in place, you'll be able to focus on what matters most: crafting a stunning underwater world that's both visually captivating and biologically sound.

CHAPTER 04

CHOOSING **THE RIGHT** PLANTS **AND** FISH FOR YOUR **AQUASCAPE**

"Selecting plants and fish that complement each other is crucial for balance in your aquascape. Fish should not disturb the plants, and plants should provide shelter and natural habitat for the fish."

Takashi Amano

When you first imagine your ideal aquascape, you likely picture a harmonious blend of lush greenery and vibrant fish swimming gracefully through the scene. Achieving that perfect balance between plants and fish is at the heart of successful aquascaping. But it's not just about picking plants that look good and fish that catch the eye — you need to choose species that thrive together, complement each other's needs, and enhance the ecosystem you're creating.

Whether you're a beginner or an experienced aquascaper, selecting the right combination of plants and fish can be a challenge. How do you know which plants are beginner-friendly and which ones need more advanced care? What fish are compatible with your plants, and how do their behaviors affect your aquascape's environment? These are critical questions that influence the health and beauty of your tank.

In this chapter, we'll dive deep into the process of selecting the right plants and fish for your aquascape. We'll break down everything from beginner-friendly plants to high-tech setups, explain the different categories of plants based on their positioning in the tank, and discuss how to balance the needs of your plants and fish. By the end, you'll be equipped with the knowledge to create a thriving, balanced aquascape that not only looks stunning but also functions as a healthy ecosystem.

Selecting Aquatic Plants

Choosing the right plants is the foundation of any aquascape. Plants provide oxygen, help control algae, and create a natural habitat for your fish. However, different plants have different care requirements, and understanding these can help you create a visually appealing and balanced tank.

Beginner-Friendly Plants

Beginner-friendly plants are generally tolerant of low light and do not require complex CO2 systems or constant fertilization. If you're just starting out, it's important to pick plants that are easy to care for, resilient in a variety of conditions, and forgiving of common mistakes. The following plants are widely regarded as some of the best options for beginners due to their hardiness and low maintenance needs.

Anubias

Anubias is a versatile, low-light plant that can thrive in a variety of conditions. Its broad, dark green leaves make it an attractive addition to any aquascape, and it grows slowly, which means less trimming and maintenance. Anubias can be attached to driftwood or rocks, and because it doesn't require substrate to grow, it can be placed in different areas of the tank for maximum visual impact.

Java Fern

Java Fern is another beginner favorite. This plant is highly adaptable and can grow in low light without the need for CO2 injection, making it perfect for low-tech tanks. It thrives when attached to hardscape materials like rocks or driftwood. Its unique leaf structure adds texture to your aquascape, and because it's a slow grower, it's ideal for hobbyists who want a low-maintenance setup.

Cryptocoryne

Cryptocoryne plants, often referred to as "Crypts," are perfect for beginners looking to add variety to their aquascape. They come in a range of colors, from deep green to reddish-brown, and can thrive in low-light conditions. Crypts are typically planted in the substrate and are known for being relatively slow growers, which makes them easy to manage. One thing to note is that they can experience "crypt melt," where the leaves temporarily die off when they're first introduced to a new tank, but they usually bounce back stronger.

Java Moss

Java Moss is one of the easiest plants to care for in a freshwater aquarium, making it ideal for beginners. It requires minimal light and no CO_2, growing well even in low-tech setups. You can attach Java Moss to rocks, driftwood, or let it float freely, creating a natural, mossy look in your aquascape. It's also great for breeding tanks, as it provides shelter for fish fry.

Hornwort

Hornwort is a fast-growing, floating plant that doesn't require any substrate, making it a versatile option for beginners. It helps to reduce nitrate levels in the tank and can thrive in a wide range of water conditions. Its dense, feathery appearance adds texture and provides hiding spots for fish. Hornwort can grow quickly, so regular trimming may be necessary.

Amazon Sword

The Amazon Sword is a classic aquarium plant that's easy to care for, even for beginners. It thrives in moderate to low light and can grow quite large, becoming a central focal point in your tank. While it benefits from root tabs for nutrition, it doesn't demand CO2 injection or complex care, making it perfect for those just starting out.

Water Wisteria

Water Wisteria is another excellent beginner-friendly plant. It can grow in a variety of conditions, including low light, and can be planted in the substrate or left to float. This plant grows relatively fast, providing excellent cover and helping to absorb excess nutrients in the water, which helps prevent algae growth. Its lacy leaves add a delicate look to your aquarium.

High-Tech Plants for Advanced Setups

Once you've mastered the basics, you might want to explore more challenging plants that require specific conditions to thrive but yield stunning results. These plants often require intense lighting, CO2 supplementation, and regular fertilization to thrive.

Monte Carlo

Monte Carlo is a popular choice for creating a lush, green carpet in high-tech aquascapes. It requires intense lighting and regular CO2 supplementation to grow properly. Monte Carlo spreads quickly when given the right conditions, forming a dense mat across the substrate, which is ideal for creating a seamless foreground.

Glossostigma

Another carpeting plant, Glossostigma elatinoides (often called "Glosso"), is a favorite among experienced aquascapers. It requires high light, CO2 injection, and a nutrient-rich substrate to thrive. Glosso grows rapidly, and its small, vibrant green leaves create a clean, uniform foreground. However, it can quickly become overgrown if not properly maintained, so regular trimming is essential.

Rotala Rotundifolia

Rotala Rotundifolia is an advanced plant that adds vibrant color to your aquascape with its pinkish-red leaves. This plant thrives in high-light conditions and requires regular CO2 injections for optimal growth. Rotala Rotundifolia grows quickly and needs frequent trimming to prevent it from becoming too dense. It's often used as a background plant to create depth and contrast in high-tech setups.

Ludwigia Super Red

Ludwigia Super Red is known for its striking red coloration, making it a popular choice for high-tech aquascapes. This plant requires intense lighting and CO2 supplementation to maintain its vibrant color and grow properly. It can grow both submerged and emersed, and its contrasting hues provide a stunning visual impact in the midground or background of the aquarium.

Bucephalandra

Bucephalandra is a relatively slow-growing plant that thrives in high-tech setups. It needs strong lighting and CO2 supplementation to achieve its best form and coloration. Bucephalandra comes in a variety of leaf shapes and colors, often showing hues of green, blue, or even purple under optimal conditions. It can be attached to rocks or driftwood, making it perfect for adding texture and interest to an aquascape.

Alternanthera Reineckii

Alternanthera Reineckii is a striking plant with deep red and purple leaves, making it a favorite among aquascapers looking for color contrast. It requires high light and CO2 to grow well, and under the right conditions, it can add a dramatic pop of color to your tank. Alternanthera Reineckii grows tall, so it's often used as a background or midground plant to create layers in an aquascape.

Hygrophila Pinnatifida

Hygrophila Pinnatifida is a unique, visually appealing plant that features serrated leaves with shades of green, brown, and red. It thrives under high light and CO2 injection and does best when attached to hardscape like rocks or driftwood. Its unusual texture and color make it an excellent choice for advanced aquascapes looking for variety in plant forms.

Staurogyne Repens

Staurogyne Repens is a compact, green plant that requires high lighting and CO2 supplementation to flourish. It is often used as a midground or foreground plant due to its low-growing, bushy nature. This plant adds structure and density to an aquascape, and with proper care, it can spread to form a dense, green carpet over time.

Categories of Plants

To create depth and balance in your aquascape, it's essential to understand the different categories of plants and how to use them effectively in your aquascape.

Background Plants
Background plants are typically taller species that fill the rear of the tank, providing structure and a backdrop for the midground and foreground plants. Common choices for background plants include Vallisneria, which grows long, ribbon-like leaves that sway beautifully in the current, and Hygrophila, a fast-growing stem plant that adds height and volume.

Midground Plants
Midground plants are medium-sized species that add depth to the center of your aquascape. These plants help transition between the tall background plants and the low-growing foreground plants. Popular midground plants include Java Fern and Cryptocoryne species, both of which offer a range of textures and colors that blend well with other plants.

Foreground Plants
Foreground plants are low-growing species that spread across the substrate to form a carpet or cover. They're often used to create an open, spacious look at the front of the tank. Carpeting plants like Monte Carlo and Dwarf Hairgrass are common choices for foreground plants, as they grow densely and remain short.

Floating Plants
Floating plants, like Water Lettuce and Amazon Frogbit, float on the surface of the water and can create an attractive, natural canopy for your tank. They provide shade, help reduce algae by blocking excess light, and offer a place for fish to hide. However, floating plants can sometimes block light from reaching the plants below, so they need to be managed carefully.

Growth Rates, Colors, and Textures: How to Combine Plants Aesthetically
A visually pleasing aquascape relies on the harmonious combination of plants with different growth rates, colors, and textures. Fast-growing plants like Vallisneria can help fill in spaces quickly while slower-growing species like Anubias maintain their structure for longer periods without frequent pruning.

When selecting plants, think about the color contrast between different species. Mixing dark green plants like Java Fern with brighter species like Ludwigia or red varieties of Cryptocoryne adds visual interest and prevents the aquascape from looking monotonous.

Textures also play a vital role in creating depth. Combine smooth-leaved plants like Amazon Swords with spiky species like Java Moss to add variety to your aquascape. The key is to balance these elements without overwhelming the viewer. Too many species can make the tank look cluttered, so aim for cohesion by limiting the number of different plant types and focusing on how they complement each other.

Low Light vs. High Light Plants

Light requirements vary widely among aquatic plants. Choosing the right plants for your lighting setup is essential for ensuring their health and growth. Understanding whether a plant thrives in low light or high light will help you create a balanced and thriving aquascape.

Low Light Plants

Low-light plants are perfect for tanks with minimal lighting or for hobbyists who prefer a low-maintenance setup. These plants grow slowly and don't require intense light or CO2 supplementation, making them ideal for beginners or low-tech tanks. Here are ten popular low-light plants:

- Java Moss
- Anubias
- Cryptocoryne
- Java Fern
- Hornwort
- Marimo Moss Ball
- Amazon Sword
- Vallisneria
- Water Wisteria
- Buccophalandra

High Light Plants

High-light plants require strong lighting and CO2 injection to thrive. These plants grow rapidly and produce more vibrant colors, but they also demand more attention in terms of trimming, fertilization, and water parameter monitoring. High-light plants are best suited for advanced aquascapers who are looking for a dynamic, fast-growing aquascape. Here are ten popular high-light plants:

- Glossostigma Elatinoides
- Rotala Rotundifolia
- ·Dwarf Hairgrass
- Monte Carlo
- Alternanthera Reineckii
- Ludwigia Super Red
- Hemianthus Callitrichoides (Cuba)
- Staurogyne Repens
- Hygrophila Pinnatifida
- Proserpinaca Palustris (Mermaid Weed)

Choosing plants based on their light requirements helps ensure they thrive in your aquarium and reduces the likelihood of growth issues.

Choosing the Right Fish for Your Aquascape

Once you've selected your plants, the next step is choosing fish that complement the aquascape both aesthetically and ecologically. The fish you select should not only enhance the beauty of your tank but also coexist peacefully with your plants and other species.

Fish Behavior and Compatibility: Community Fish vs. Species-Only Tanks

Fish behavior plays a significant role in the success of your aquascape. Some fish are peaceful and can live in harmony with a variety of species, while others may be territorial or aggressive. Understanding the behavior of different fish species is crucial for maintaining a balanced tank.

Community Fish Tanks

Community tanks are popular because they allow you to mix different species, each bringing its own visual appeal and personality. Species like Tetras, Rasboras, and Corydoras are known for their peaceful nature, making them ideal for community setups. These fish swim in schools, creating a dynamic, flowing movement in your aquascape, which complements the plant life.

When creating a community tank, ensure that all fish species have similar water requirements in terms of temperature, pH, and hardness. It's also important to consider the size of the fish and their social behavior. Aggressive or territorial species can disrupt the peaceful balance of a community tank, leading to stress and potential harm to other fish.

Species-Only Tanks

In contrast, species-only tanks focus on a single species of fish. This approach allows you to tailor the tank to the specific needs of that species, whether it's a school of Neon Tetras or a group of Discus. Betta fish, for example, are often kept in species-only tanks due to their aggressive nature toward other fish. In a species-only setup, you can tailor the tank environment to the specific species, ensuring optimal conditions for their health and well-being.

Species-only tanks can be visually striking, especially when designed around the natural habitat of the fish. While species-only tanks may lack the diversity and dynamic interactions found in community tanks, they offer an opportunity to observe the intricate behaviors and interactions of a single species. This type of setup is particularly appealing for aquarists who enjoy breeding fish or studying their natural habits.

Popular Fish for Aquascaping

Certain fish species are particularly well-suited for aquascaping due to their size, behavior, and compatibility with plants.

Tetras

Tetras, such as Neon Tetras and Cardinal Tetras, are among the most popular choices for aquascaping. These small, colorful fish are peaceful, schooling fish that thrive in planted tanks. Their vibrant colors stand out against the greenery of the plants, creating a striking visual effect. Tetras are also relatively easy to care for, making them an excellent choice for beginners.

Rasboras

Rasboras, like Harlequin Rasboras and Chili Rasboras, are another favorite among aquascapers. These fish are small, active, and peaceful, making them ideal for community tanks. Rasboras are known for their striking colors and their tendency to school tightly, which adds movement and life to the aquascape.

Otocinclus Catfish

Otocinclus catfish, often called "Otos," are small, peaceful algae eaters that are perfect for planted tanks. They help keep the aquarium clean by feeding on algae growing on plants and glass. Otos are gentle and thrive in groups, making them a valuable addition to any aquascape that needs a bit of algae control without harming delicate plants.

Corydoras

Corydoras catfish are bottom-dwelling fish that are perfect for keeping the substrate clean. These small, peaceful fish are known for their playful behavior and are a great addition to any aquascape. Corydoras are compatible with most community fish and can help keep the tank clean by scavenging for leftover food.

Dwarf Gourami

Dwarf Gouramis add a touch of elegance and color to aquascapes with their vibrant blue, red, or orange hues. They are peaceful fish that prefer well-planted tanks with lots of hiding spots. Their graceful swimming and beautiful colors make them a popular choice for aquascapers looking to add a feature fish to the tank.

Betta Fish

Betta fish are a popular centerpiece in aquascaped tanks, thanks to their stunning colors and flowing fins. While Betta fish can be aggressive towards other Betta fish, they are generally peaceful with other community fish. Bettas thrive in heavily planted tanks, where they can explore and find hiding places.

Ember Tetras

Ember Tetras are small, bright orange fish that provide a stunning contrast against the green plants in an aquascape. They are peaceful, schooling fish that do well in community tanks. Their small size and vibrant color make them ideal for smaller aquariums or nano aquascapes.

Rummy Nose Tetras

Rummy Nose Tetras are another excellent schooling fish known for their striking red noses and silver bodies. They add a lot of movement to the tank as they swim together in groups, creating a dynamic visual effect in the aquascape. They are peaceful and pair well with other community fish.

Pearl Gourami

Pearl Gouramis are known for their delicate, pearly sheen and long, flowing fins. They thrive in well-planted tanks and enjoy hiding in vegetation. Their calm, peaceful nature and beautiful appearance make them a great choice for aquascapes, especially for hobbyists looking for a larger feature fish.

Endler's Livebearers

Endler's Livebearers are small, brightly colored fish that are easy to care for and breed. They are hardy and adaptable to various tank conditions, making them suitable for beginner and advanced aquascapers alike. Their bright colors provide a nice contrast to the green plants, and their active nature adds life to the tank.

Blue Rams

Blue Rams, or German Blue Rams, are small cichlids known for their vivid blue and yellow coloration. They are peaceful and do well in planted tanks with soft water. Their striking colors make them stand out in an aquascape, and they add personality and beauty to the aquarium.

Hatchetfish

Hatchetfish are unique surface-dwelling fish that add an interesting dynamic to aquascaped tanks. Their sleek, silvery bodies and habit of swimming near the water's surface provide contrast to the mid and bottom-dwelling species. They are peaceful and do well in community tanks with lots of floating plants for cover.

Fish That Complement Plants: Herbivores vs. Carnivores

When selecting fish for your aquascape, it's important to consider their feeding habits and how they might interact with your plants. Herbivores can help manage algae, while carnivores can control small invertebrates without disturbing your plants.

Herbivores

Herbivorous fish are often beneficial for planted tanks, as they help control algae growth by grazing on them. However, some may also nibble on plants, so it's essential to choose species that won't damage your aquascape. Here are 10 popular herbivorous fish:

- Siamese Algae Eater
- Otocinclus Catfish
- Amano Shrimp (though not a fish, highly effective for algae control)
- Plecostomus (Bristlenose variety is plant-safe)
- Silver Dollar Fish
- Mollies
- Hillstream Loach
- Rosy Barb
- Cherry Shrimp
- Florida Flagfish

Carnivores

Carnivorous fish primarily eat insects, larvae, and small invertebrates. They are less likely to disturb plants but may prey on smaller fish or shrimp. These fish are ideal for tanks where you want to avoid plant damage. Here are 10 popular carnivorous fish:

- Betta Fish
- Dwarf Gourami
- Angelfish
- Pea Puffer
- German Blue Ram
- Apistogramma
- Killifish
- Electric Blue Acara
- Clown Loach
- Pictus Catfish

By carefully selecting herbivorous and carnivorous fish, you can create a harmonious aquascape where plants and fish thrive together.

Balancing Plants and Fish

A successful aquascape is one where plants and fish live in harmony, each contributing to the health of the overall ecosystem. Balancing the needs of both plants and fish is essential for maintaining a thriving tank.

Managing Fish Waste for Plant Nutrition

Fish waste is a natural source of nutrients for plants, providing essential elements like nitrogen and phosphorus. In a balanced aquascape, fish waste can help fertilize the plants, reducing the need for chemical fertilizers. However, too much waste can lead to an excess of nutrients, which can cause algae blooms and water quality issues.

It's important to choose the right number of fish for your tank's size and the plants' nutrient needs to manage fish waste effectively. Regular water changes, filtration, and the use of substrate fertilizers can help maintain the balance between fish waste and plant nutrition.

Creating Harmony Between Plants and Fish: Avoiding Common Issues

Common issues in aquascapes often arise from imbalances between the plants and fish. For example, overstocking your tank with fish can lead to excessive waste, which overwhelms the plants and promotes algae growth. On the other hand, understocking the tank may result in nutrient deficiencies, as the plants don't receive enough waste to support their growth.

To avoid these issues, it's important to research the specific needs of both your plants and fish. Pay attention to factors like water parameters, lighting, and feeding habits to ensure that both your plants and fish are thriving. Regular monitoring of water quality, trimming plants, and feeding your fish the right amount will help create a harmonious environment where both plants and fish can flourish.

The Bottom Line

Choosing the right plants and fish for your aquascape is a delicate balancing act, but with careful planning and attention to detail, you can create a thriving underwater ecosystem that is both beautiful and sustainable. By selecting beginner-friendly plants or more advanced species based on your setup, you can ensure that your tank's plants will grow healthy and strong. Pairing the right fish with those plants — whether peaceful Tetras, active Corydoras, or algae-eating Otocinclus — will enhance the overall aesthetic and health of your aquascape.
Remember, creating harmony between your plants and fish is the key to a successful aquascape. When both thrive together, your tank will not only look stunning but will also provide a healthy, balanced environment for all its inhabitants.

CHAPTER
05

DESIGN **PRINCIPLES AND** TECHNIQUES FOR **AQUASCAPING**

"The most important principle in aquascaping is balance. The arrangement of plants, rocks, and wood should create harmony, reflecting the beauty of nature."

Takashi Amano

Aquascaping is often described as the art of creating an underwater landscape. But beyond the stunning visuals, it's an art form that balances aesthetic principles with the needs of a living ecosystem. Whether you're crafting a serene, minimalist design or a lush, jungle-like scene, aquascaping is about using natural elements to create a world that feels both familiar and otherworldly. This chapter dives into the essential design principles and techniques that help transform a basic aquarium into a vibrant, living work of art.

If you've ever been captivated by an aquascape, it's likely because of how all the elements—plants, rocks, water, and fish—work together to form a cohesive, harmonious environment. But creating this kind of beauty requires more than just picking the right plants or fish. It's about understanding the fundamental design principles that guide artists and landscape designers alike: balance, contrast, depth, and focus. For aquascapers, mastering these principles means not only elevating the visual appeal of the aquarium but also ensuring the health and sustainability of the plants and fish within.

In this chapter, you'll learn how to apply these design principles to your aquascape. We'll explore aesthetic concepts like the rule of thirds and the golden ratio, dig into layout styles that give structure to your designs, and discuss how to use hardscape materials to create a solid foundation. You'll also discover how planting techniques, color contrast, and focal points can transform your aquarium from a simple tank into a breathtaking, natural masterpiece. By the end, you'll have a comprehensive understanding of how to design your aquascape with intention, creativity, and technical skills.

Aquascaping as Art: Aesthetic Concepts

Aquascaping, often described as the art of underwater gardening, is much more than just arranging plants and rocks. It's about creating a harmonious, balanced composition that captivates the viewer and transports them into a miniature world of natural beauty. When done thoughtfully, your aquascape becomes a living piece of art, blending form and function to support aquatic life while providing an aesthetically pleasing scene. Much like traditional art forms, aquascaping relies on certain design principles to ensure that the final result feels cohesive and visually engaging. In this subtopic, we'll explore some essential aesthetic concepts in aquascaping, such as the rule of thirds, the golden ratio, and techniques for creating depth and perspective.

The Rule of Thirds

One of the foundational principles in aquascaping is the rule of thirds, a concept borrowed from photography and painting. This rule helps you structure your aquarium's layout to achieve a balanced and pleasing composition. To apply the rule of thirds, imagine dividing your aquarium into nine equal parts by using two horizontal and two vertical lines. By placing key elements, like focal plants, rocks, or driftwood, along these lines or at their intersections, you naturally guide the viewer's eye across the aquascape.

For instance, if you're using a large rock formation as a focal point, placing it off-center along one of the vertical lines can make the tank feel more dynamic. Similarly, if you have a bright-colored plant, positioning it at one of the intersections of the grid ensures it draws attention without overwhelming the viewer. Research in visual arts has shown that compositions adhering to the rule of thirds are often perceived as more balanced and natural. This principle can take your aquascaping to the next level by creating a flow that feels intuitive to the viewer.

The Golden Ratio

Closely related to the rule of thirds is the golden ratio, a mathematical concept that has been used in art, architecture, and design for centuries. The golden ratio, approximately 1:1.618, is a proportion that is often found in nature, such as in the spirals of shells or the branching of trees. In aquascaping, the golden ratio can help guide the placement of hardscape elements, plants, and even open spaces to create a design that feels organic and visually harmonious.

Takashi Amano, a pioneer in nature-style aquascaping, frequently used the golden ratio in his layouts to mimic the natural beauty of landscapes. By positioning the main focal points along these proportions, Amano's aquascapes conveyed a sense of effortless beauty and balance. For example, placing a prominent piece of driftwood or a cluster of rocks at the golden ratio line can anchor the composition, creating a flow that feels both intentional and natural.

Creating Depth and Perspective in Your Aquascape

One of the biggest challenges in aquascaping is creating the illusion of depth in what is essentially a two-dimensional space. The goal is to make the aquarium feel expansive, as though you're looking into a real, living landscape. To achieve this, there are several key techniques you can employ. First, consider the size and placement of plants and hardscape materials. Using smaller plants and rocks in the foreground and larger ones toward the back helps create a sense of depth. For example, by planting Dwarf Hairgrass or Eleocharis acicularis in the foreground and using taller plants like Cryptocoryne or Vallisneria in the background, you can guide the viewer's eye deeper into the aquascape. This technique mimics the way we perceive depth in nature, where distant objects appear smaller.

Layering is another powerful tool for enhancing depth. By selecting plants with different heights, textures, and leaf shapes, you can build visual layers within the tank. For instance, using fine-leaved plants in the middle ground, such as Rotala rotundifolia, and broader-leaved plants, like Anubias, in the background creates visual interest. This layered approach makes the aquarium feel more three-dimensional and immersive.

Additionally, the placement of hardscape elements like rocks and driftwood can help direct the viewer's gaze. The lines and curves of a piece of driftwood, for example, can naturally lead the eye from the front of the tank to the back, enhancing the illusion of depth. Rocks can be arranged to suggest natural formations, such as cliffs or riverbanks, creating a more dynamic and engaging scene.

Lighting also plays a crucial role in creating depth. By adjusting the intensity and direction of light, you can highlight certain areas of the tank while allowing others to recede into shadow. For instance, spotlighting a key plant or rock formation in the foreground while keeping the background dim can add dramatic contrast, making the tank feel deeper and more layered.

Finally, playing with substrate levels can significantly enhance the sense of depth in your aquascape. By raising the substrate at the back of the tank or creating slopes, you give the illusion of uneven terrain, similar to what you'd find in a natural environment.

This technique adds visual interest and gives the aquarium a more three-dimensional feel.

Basic Layout Styles

When it comes to aquascaping, your layout is the foundation that dictates the overall look and feel of your aquarium. Think of it like the framework for a painting, where the composition is carefully planned to create a visually engaging and harmonious design. In aquascaping, this means choosing how plants, rocks, and other materials are arranged to bring your vision to life. Each layout has distinct characteristics that suit different styles and purposes. By understanding these basic layout styles, you'll be able to select the one that aligns with your goals for your freshwater aquarium.

Triangular Layout

One of the most common and visually striking styles is the triangular layout. This design introduces a strong diagonal line that stretches from one corner of the tank to the opposite side. Imagine tall plants or rocks forming a peak on one end and then tapering down as you move across the tank. This creates a dynamic, asymmetrical look that naturally guides the viewer's eye from one side of the aquarium to the other.

For example, let's say you want to evoke the sense of a riverbank in your aquascape. You might place larger rocks and tall plants like Vallisneria or Java Fern on one side, gradually leading to smaller, more delicate plants on the opposite end. This technique creates a sense of movement, like the flow of water, while still maintaining balance. The triangular layout works especially well in aquariums where you want to make a strong visual impact without overcomplicating the design.

U-Shape Layout

The U-shape layout is another popular choice, known for creating a concave or bowl-like appearance. In this layout, taller plants or rocks are placed along both sides of the tank, while the middle remains open or contains low-growing plants. This creates a natural focal point in the center of the aquarium, drawing attention to a specific feature such as a striking piece of driftwood or a vibrant plant.

Imagine a 40-gallon tank where you place dense clusters of taller plants like Amazon Sword along the edges. In the center, you could leave space for a low-growing carpet of Dwarf Baby Tears. This creates an inviting, serene scene with a central focus, making the tank feel deeper and more expansive. The U-shape layout is perfect for aquascapers who want a balanced, yet visually intriguing design.

Island Layout

If you prefer simplicity with a touch of elegance, the island layout might be the perfect fit. This style revolves around placing a central "island" of plants, rocks, or driftwood in the middle of the tank, with open space surrounding it. The island layout creates a minimalist look, where the central element becomes the star of the show.

Concave Design

A concave design takes the U-shape layout one step further, emphasizing the hollow or bowl-like appearance even more. By raising the plants or rocks along the sides and creating a dip in the middle, this layout creates a peaceful, immersive environment that mimics the feeling of being in an enclosed underwater valley. The design draws viewers into the tank, creating a sense of depth and tranquility.

You can enhance the concave design by using plants of varying heights along the sides, such as Cryptocoryne and Rotala, while keeping the center lower with carpeting plants like Monte Carlo. This layout works particularly well in aquariums where you want to create a more intimate and enclosed space, perfect for a peaceful aquascape that feels immersive.

Convex Design

Conversely, the convex design takes a different approach by pushing elements toward the center of the tank. In this layout, plants or rocks rise higher in the middle and gradually slope downward toward the sides. This creates a bold central focal point, often evoking the natural look of hills or mountain ranges.

For example, in a 55-gallon aquarium, you might build up a mound in the center with rocks and higher-growing plants like Ludwigia. As the plants taper off toward the sides, you create the illusion of a hill rising from the tank's base. This dynamic design draws attention to the center, making it a great choice for aquascapers looking to create a more dramatic, nature-inspired look.

Which Style to Choose for Your Aquarium

Choosing a layout style for your aquarium depends on the aesthetic and experience you want to create. If you're drawn to dynamic, asymmetrical designs, a Triangular Layout could be ideal, with taller plants or rocks tapering down across the tank. For a serene, balanced look, the U-Shape Layout provides a natural focal point, often in the center. The Island Layout offers simplicity and elegance, perfect for smaller tanks, while Concave and Convex Designs create depth and drama, mimicking valleys or hills. Select the style that aligns with your vision to bring your aquascape to life.

Using Hardscape to Create Structure

Hardscape elements like rocks, driftwood, and substrates form the backbone of your aquascape. These materials provide structure and depth, shaping the layout and guiding the placement of plants. Thoughtfully placed hardscape can transform a flat, lifeless tank into a dynamic, natural environment.

Placing Rocks, Driftwood, and Substrates to Build a Strong Foundation

When placing hardscape materials, think about how they interact with each other and the overall design. Rocks, for example, should be arranged in a way that mimics natural formations. Instead of placing rocks evenly across the tank, try clustering them together in one area, creating a sense of movement and flow. Use rocks of varying sizes to add depth and perspective, with larger rocks toward the back and smaller ones in the foreground.

Driftwood is another essential element for adding texture and structure to your aquascape. Whether you're going for a minimalist look or a more intricate design, driftwood can create a sense of organic movement, particularly when placed diagonally or at an angle. Just like with rocks, avoid placing driftwood in straight lines. Instead, use it to guide the eye across the tank or to form natural divisions between different areas of the aquascape.

Substrate choice is equally important, as it provides both a visual base and a practical one for anchoring plants. Fine gravel or sand can be used to create smooth, even surfaces, while larger, chunkier substrates add texture. You can also use substrate to build up areas of elevation, creating hills or valleys that enhance the depth of your aquascape.

Planting Techniques

Once you've established the hardscape, it's time to add plants. The way you plant your aquarium can greatly affect its final appearance, so careful attention should be paid to where and how each plant is placed.

Anchoring Plants in the Substrate

Anchoring plants in the substrate is the most common planting technique. For plants with roots, such as Cryptocorynes or Amazon Swords, you'll want to bury the roots deep enough to secure the plant without covering the crown (the part where the roots meet the leaves). This prevents rot and encourages healthy growth. Use planting tweezers to help position the plants precisely without disturbing the substrate.

When arranging plants, consider how they will grow over time. Place faster-growing species in areas where they have room to spread and leave space between plants to allow for future growth. If you're using carpeting plants, such as Monte Carlo or Glossostigma, plant them in small clumps spread evenly across the substrate to encourage them to fill in the gaps.

Attaching Plants to Rocks and Driftwood

Some plants, like Anubias and Java Fern, don't need to be anchored in the substrate. Instead, they can be attached directly to rocks or driftwood. These plants have rhizomes, which should not be buried, as doing so can cause the plant to rot. Instead, use plant-safe glue or fishing line to attach the plant to your hardscape.

Attaching plants to rocks and driftwood adds an extra layer of depth and texture to your aquascape. The plants will grow along the contours of the hardscape, softening the edges of rocks and creating a more natural, cohesive look. Over time, these plants will root themselves securely into the hardscape, blending seamlessly into the design.

Layering and Grouping Plants for Visual Impact

Grouping plants by height, color, and texture is key to creating a visually appealing aquascape. Tall plants, like Vallisneria or Rotala, are best suited for the background, where they can create a lush wall of greenery. Mid-sized plants, such as Cryptocorynes or Swords, work well in the midground, adding volume without overpowering the scene. Foreground plants, like Monte Carlo or Dwarf Hairgrass, provide a soft, low-lying carpet that frames the front of the tank.

When layering plants, think about how their colors and textures complement each other. For example, pairing smooth-leaved plants with spiky or fern-like species can create contrast and prevent the tank from looking too uniform. Similarly, using plants with different growth rates will add variation to the design, as faster-growing species will fill in space more quickly while slower-growing plants provide long-term structure.

Using Color and Contrast

The use of color and contrast is one of the most effective ways to create a dynamic aquascape. By combining plants with different colors, textures, and leaf sizes, you can draw attention to specific areas of the tank and create visual interest.

Combining Different Textures, Colors, and Leaf Sizes

Mixing textures and leaf shapes is an excellent way to add variety to your aquascape. Broad-leaved plants, like Amazon Swords, create a bold, structured look, while fine-leaved plants, such as Cabomba or Myriophyllum, add a delicate, flowing quality to the tank. Combining these textures will prevent your aquascape from feeling flat or monotonous.

Color is another important factor. Green is the dominant color in most aquascapes, but by introducing plants with red, purple, or even brown leaves, you can create striking contrasts. Red plants, like Ludwigia or Rotala, work especially well as focal points, standing out against the green backdrop of other plants.

Adding Red Plants for Contrast: How to Manage Light and Nutrients

Red plants add vibrant color to any aquascape, but they require specific conditions to thrive. Most red plants need high-intensity light and a steady supply of nutrients, particularly iron, to maintain their color. Without these conditions, red plants can lose their vibrancy and turn green.

CO_2 supplementation is also crucial for red plants, as it encourages faster growth and helps the plants absorb the nutrients they need. If you're planning to include red plants in your aquascape, be prepared to invest in a high-tech setup with strong lighting and regular fertilization.

Creating Focal Points

Focal points play a central role in designing a captivating aquascape. They are the elements that immediately grab your attention when you glance at the aquarium. These focal points can range from a striking piece of driftwood to a vibrant, colorful plant. In the broader context of aquascaping, creating focal points is key to guiding the viewer's eye and establishing balance within the design. Without these elements, an aquascape might feel flat or lacking direction. By thoughtfully incorporating focal points, you can elevate your tank's aesthetic and keep it engaging.

Using Hardscape and Plants to Draw Attention

When you set out to design your aquascape, one of your main tasks is to determine where you want the viewer's attention to land. This is where focal points come into play. Hardscape elements, such as rocks, driftwood, or even decorative stones, provide a great foundation for these focal points. For instance, a large, textured piece of driftwood can act as a visual anchor, giving the eye a place to rest before exploring the rest of the tank. Positioning this driftwood in accordance with design principles like the rule of thirds or the golden ratio will ensure that the focal point feels natural and balanced, not forced.

To bring this to life, imagine you have a 50-gallon tank. You can place a prominent rock formation off-center, aligning it along the lines of the rule of thirds. The viewer's eye will naturally be drawn to this structure, making it the heart of your design. But don't stop there—complement the hardscape with plants. Adding soft mosses around the base of the rocks or placing taller plants behind the formation will create depth and add more visual layers.

Plants as Focal Points

Plants can also be excellent focal points, especially when they offer a stark contrast to their surroundings. For example, a bright red plant like Alternanthera Reineckii stands out beautifully against a backdrop of lush green foliage. This contrast helps to create a strong visual emphasis, pulling attention toward that specific area of the tank.

Let's say you're working with a tank that primarily features green plants like Java Fern or Anubias. Introducing a single bold red or orange plant in the center will create a striking focal point, adding variety to the color palette and increasing visual interest. This technique not only makes the aquascape more vibrant but also introduces a sense of depth as the eye naturally moves from the bold focal point to the softer surrounding greenery.

Balancing Focal Points for Harmony

It's important to remember that focal points work best when they're balanced with the rest of the aquascape. If you overload your tank with too many attention-grabbing features, the design can feel chaotic rather than harmonious. Think of focal points as the highlights of your composition, not the entire picture. By spacing them out and using other design elements like plants and stones to frame or support them, you create a more cohesive and dynamic layout.

For example, if you've placed a tall, spindly piece of driftwood as your main focal point, balance it by placing lower-growing plants around it. You can also introduce a smaller secondary focal point—such as a small rock formation on the opposite side of the tank—to create balance without overwhelming the design.

By thoughtfully arranging focal points, you control how the viewer experiences your aquascape. The interplay between hardscape, plants, and placement creates a dynamic flow that keeps the design engaging and visually appealing.

The Bottom Line

Designing a stunning aquascape requires more than just a basic understanding of plants and fish. It's about applying artistic principles—like balance, contrast, and perspective—while also working with the unique needs of a living ecosystem. Whether you're experimenting with different layout styles, using hardscape to create structure, or selecting plants for their color and texture, each decision contributes to the overall harmony of the aquarium.

As you continue your aquascaping journey, remember that there's no one right way to create a beautiful tank. The best aquascapes reflect the unique vision of the aquascaper, combining creativity with an understanding of nature. With practice, patience, and a willingness to experiment, you'll find your own approach to designing an aquascape that not only looks stunning but also thrives in a healthy, balanced ecosystem.

CHAPTER 06

SETTING **UP** YOUR **AQUASCAPE** **STEP**-BY-**STEP**

"Patience is key when setting up an aquascape. Take your time with the layout, as this will form the backbone of your entire tank. Rework it until you're happy before adding water or plants."

George Farmer

Embarking on the journey of creating a captivating aquascape can feel overwhelming at first, but it's also an exciting opportunity to craft your very own underwater world. Whether you're drawn to the serene beauty of a planted tank or the dramatic contrast of rocks and driftwood, the setup process is where everything starts to come together. This chapter will guide you through the process of designing and building your aquascape step by step, ensuring that each stage—from sketching your vision to cycling your aquarium—runs smoothly.

The key to a successful aquascape is thoughtful planning and careful execution. Without a clear roadmap, it's easy to run into issues like poor plant growth, unbalanced ecosystems, or water clarity problems. But with the right approach, you'll not only achieve a stunning design but also create a healthy environment for your plants and fish to thrive. This chapter will help you make informed decisions, prepare all necessary materials, and avoid common pitfalls along the way.

By following this guide, you'll gain the confidence to bring your aquatic vision to life and maintain it for years to come. Let's dive into the step-by-step process of setting up your aquascape.

Pre-Setup Planning

Setting up an aquascape is an exciting process, but it requires careful planning. Before you even touch your aquarium or buy your first plant, you need to have a clear plan. Rushing into an aquascape without thinking it through can lead to poor design choices, wasted money, and, in some cases, a failed ecosystem. Planning your aquascape gives you direction and ensures that you have all the materials and equipment needed to set up your aquarium properly. In this section, we'll dive into the importance of pre-setup planning, highlighting how having a solid strategy from the beginning sets the stage for a successful and visually stunning aquascape.

Sketching Your Design

A well-planned aquascape begins with a clear vision, and there's no better way to solidify your ideas than by sketching out your design. While it may be tempting to dive in without a plan, sketching helps you see the bigger picture and make better decisions. Think of your sketch as a roadmap that will guide you through each stage of the setup process. You don't need to be an artist—your sketch is just a tool for visualizing the placement of key elements like plants, rocks, and driftwood.

Start by deciding on a basic layout style. Are you leaning toward a triangular design, where taller elements are clustered on one side and taper toward the other? Or perhaps you're envisioning an island layout with a central focus surrounded by open space? Sketching different options allows you to play with these ideas, helping you avoid common mistakes like overcrowding or awkward placement.

Let's say you're drawn to a concave design, which creates a dip in the center of the tank while the sides are elevated. By sketching this out, you can plan how to distribute plants of varying heights, ensuring that taller plants frame the sides and shorter ones fill the middle. This helps achieve balance and harmony in the aquascape. Using design principles like the rule of thirds or golden ratio while sketching can further enhance the visual flow and make your aquarium more aesthetically pleasing.

Checklist of Materials and Equipment

After finalizing your design sketch, it's time to gather the materials and equipment needed to bring your vision to life. An aquascape is more than just an aquarium and plants—it's a living ecosystem that requires the right tools to thrive. Having a comprehensive checklist of everything you'll need ensures that you're fully prepared, preventing any last-minute trips to the store or interruptions in the setup process. Here's a basic checklist of essentials:

- **Aquarium Tank:** Select a tank size that complements your design and fits the available space. Larger tanks provide more room for complex designs, but smaller tanks can still be stunning with thoughtful planning.
- **Substrate:** The substrate is the foundation of your aquascape, providing both structural support and nutrients for your plants. Choose between nutrient-rich substrates, which help promote healthy plant growth, or inert substrates, which are better for setups with fewer plants.
- **Hardscape Materials:** Rocks, driftwood, and other decorative elements are essential for creating structure and depth. For example, large pieces of driftwood can serve as focal points, while smaller stones can help frame and accentuate specific areas.

- **Plants:** Plants come in various types—foreground, midground, and background. Foreground plants, such as Dwarf Hairgrass, are shorter and fill in the front, while midground plants like Cryptocoryne provide balance in the center. Background plants, such as Vallisneria, are taller and add height to the back of the tank.
- **Lighting System:** Lighting is critical to plant growth. LED lights are energy-efficient and versatile, while fluorescent lights can also support plant growth. Consider the specific lighting needs of the plants you've chosen—some require more intense light than others.
- **Filtration System:** A good filtration system is vital for maintaining water quality. Depending on your tank size and design, you can opt for an internal, external, or sponge filter. Larger tanks often benefit from external filters, while smaller setups may work well with internal or sponge options.
- **CO2 System:** While not always necessary, a CO2 system can enhance plant growth, especially in high-tech aquascapes. Plants need carbon dioxide to thrive, and adding a CO2 system can help them grow faster and healthier.
- **Aquarium Heater:** Depending on the type of fish you plan to keep, a heater may be required to maintain stable water temperatures. This is especially important for tropical species that need warmer environments.
- **Fertilizers:** Fertilizers provide plants with essential nutrients. Liquid fertilizers are added directly to the water, while root tabs are inserted into the substrate. Choose based on the needs of your specific plants.
- **Thermometer:** Monitoring water temperature is crucial to ensure the health of your tank's inhabitants. A simple thermometer can help you keep track of any fluctuations.
- **Water Conditioner:** Tap water contains chemicals like chlorine and chloramine, which can be harmful to fish and plants. A water conditioner neutralizes these chemicals, making the water safe for your tank.
- **Test Kits:** Regular water testing is necessary to monitor levels of ammonia, nitrites, nitrates, and pH. This helps prevent harmful spikes that could endanger your plants and fish.
- **Planting Tools:** Invest in aquarium-safe planting tools like scissors, tweezers, and siphons. These make it easier to trim plants, place new ones, and clean the tank without disrupting the ecosystem.

Having all these materials ready before starting the physical setup makes the process much smoother. For example, if you start arranging plants without the right lighting or filtration system in place, you risk stunting plant growth or having water quality issues. Similarly, forgetting to condition your water could harm your fish or plants.

Step-by-Step Setup

Now that you've planned your design and gathered your materials, it's time to begin the actual setup of your aquascape. This is where your vision begins to take shape, piece by piece. The following step-by-step process will guide you through preparing the tank, placing the hardscape, planting, and filling the tank with water.

Preparing the Tank and Substrate

Before placing any plants or decorations, you'll need to prepare the tank itself. Even if the tank is brand new, you'll want to rinse it out with water to remove any dust or debris. Start by cleaning the inside of the tank with water and a soft sponge—avoid using any chemicals or soap, as these can leave harmful residues that affect the ecosystem. Once cleaned, place the tank in its final location—remember, moving a filled tank is extremely difficult.

Once the tank is clean, it's time to add the substrate. If you're using a nutrient-rich substrate, place it in the tank first, ensuring an even layer across the bottom. You can create slight slopes or elevated areas to add depth to the design. Some aquascapers prefer to build the substrate higher in the back of the tank to enhance the sense of perspective, making the aquarium look deeper than it is.

If you're using a two-layer substrate system (nutrient-rich bottom layer and inert sand or gravel on top), make sure to distribute the layers evenly, preventing any mixing that could cloud the water later on. The substrate serves as both the visual base of your aquascape and a critical source of nutrients for your plants, so take care when arranging it. The depth of the substrate will depend on your plant choice—most plants need about 2-3 inches for their roots.

Placing Hardscape: Rocks, Driftwood, and Decorations

With the substrate in place, it's time to add your hardscape. This includes rocks, driftwood, or any other decorative elements that will form the backbone of your aquascape. Refer to the sketch you created earlier to guide your placement.

Start with the largest pieces of driftwood or rock, positioning them according to the design you sketched earlier. Keep in mind the rule of thirds or golden ratio to ensure that the hardscape is visually balanced.

When placing rocks, use different sizes to create a natural look. Avoid symmetrical or overly structured arrangements, as the goal is to mimic nature's organic, uneven formations. You can also use rocks to create caves or shelters for fish, adding both functionality and aesthetics to your aquascape.

For driftwood, try angling pieces diagonally or allowing them to intertwine naturally with the rocks. Driftwood provides a sense of movement and flow within the tank, especially when placed at angles rather than lying flat.

Once the larger pieces are in place, add smaller rocks or decorative elements around them to enhance the natural look. Don't forget to secure any loose elements with aquarium-safe glue or fishing line. This ensures that your design won't shift once the plants and water are added. Don't worry about perfection at this stage—once the plants are added, they will soften the edges and make everything look more cohesive.

Planting Your Aquascape: Foreground, Midground, and Background Plants

With your hardscape in place, it's time to add the plants. The key to successful planting is to think in layers: foreground, midground, and background.

Start by planting the background. Use taller species, such as Vallisneria or Rotala, to fill the back of the tank, creating a green backdrop for the rest of the plants. These plants not only frame the aquascape but also provide cover and hiding spaces for fish.

Next, move to the midground. This layer consists of medium-height plants like Cryptocoryne or Java Fern, which add volume and structure to the middle of the tank. Place them in front of the background plants, but leave enough space to avoid overcrowding. Ensure you group these plants in small clusters for a more natural look, ensuring that they have enough space to grow without overshadowing the foreground.

Finally, add the foreground plants. Low-growing species like Monte Carlo or Dwarf Hairgrass are ideal for creating a carpet-like effect at the front of the tank. Plant these in small clumps, spaced evenly apart, allowing them to spread out over time and cover the substrate. Foreground plants require more attention during planting, as they need to be spaced evenly to ensure proper coverage over time.

If you're attaching plants to rocks or driftwood, such as Anubias or Java Moss, use aquarium-safe glue or fishing line to secure them in place. Be careful not to bury rhizomes, as this can cause plants to rot.

Filling the Tank with Water: Tips for Avoiding Displacement

Now that your plants and hardscape are in place, it's time to fill the tank with water. But be careful—if water is added too quickly, it can disturb the substrate and uproot plants, ruining all the hard work you've done.

One of the most effective methods is to place a plate or shallow bowl in the tank before pouring in the water. This helps disperse the flow of water and prevents the substrate from being displaced.

Fill the tank slowly, especially in the early stages. As the water rises, you may need to adjust some of the plants or hardscape elements if they shift slightly. Once the tank is full, check to make sure that everything is still in place and that no major disturbances occurred during the fill-up.

You can also add water conditioners at this stage to remove chlorine, chloramine, or other harmful substances from tap water. It's essential to ensure that the water is safe for both plants and fish before moving on to the next steps.

Starting the Nitrogen Cycle

With the tank set up, it's tempting to add fish right away, but your aquarium needs time to establish a healthy ecosystem first. This process, known as the nitrogen cycle, is crucial to ensuring that your tank is safe for fish and other aquatic life.

What Is the Nitrogen Cycle and Why It Matters?

The nitrogen cycle is the process by which beneficial bacteria in your aquarium convert toxic ammonia, produced by fish waste and decaying organic matter, into less harmful substances like nitrites and nitrates. Without these bacteria, ammonia can build up to dangerous levels, which could poison your fish and plants. In a balanced aquarium, this cycle works to break down harmful chemicals and keep water quality stable.

To put it simply, the nitrogen cycle acts like a natural filtration system. When fish produce waste or uneaten food decomposes, ammonia is released into the water. If left unchecked, ammonia becomes toxic at even low concentrations, often leading to what's called "new tank syndrome." This can be deadly for fish if they are introduced too soon into an uncycled tank. The beneficial bacteria that colonize your filter media and substrate convert ammonia into nitrites, which are still toxic but less harmful than ammonia. Then, another type of bacteria converts nitrites into nitrates, which are much safer in small quantities. Plants can absorb some nitrates, and regular water changes help remove the excess.

By establishing the nitrogen cycle before introducing fish, you ensure that your aquarium is a stable environment where aquatic life can thrive. It may take time, but this step is critical to long-term success in aquascaping.

Cycling Your Aquarium: Step-by-Step Instructions

Cycling your tank involves creating conditions where beneficial bacteria can grow and establish themselves in the water, substrate, and filter. Fishless cycling is one of the most humane and effective methods to do this, as it allows you to establish the nitrogen cycle without harming any fish in the process. Here's a detailed step-by-step guide to get you started.

- **Add an Ammonia Source:** To kickstart the nitrogen cycle, you need to introduce a source of ammonia into the tank. One way to do this is by adding fish food, which will break down and release ammonia as it decomposes. Another method is to use pure ammonia, which is available at most pet stores. Be careful when dosing, as you only need a small amount to get the cycle going. Ammonia levels around 2-4 ppm (parts per million) are sufficient to start.
- **Monitor Water Parameters:** Over the next few days, it's essential to monitor ammonia, nitrite, and nitrate levels in your tank. Using a water testing kit, check your water daily. At first, you'll see ammonia levels rise as waste accumulates. Then, as bacteria start to develop, ammonia levels will drop, and nitrite levels will rise. This signals that the bacteria converting ammonia into nitrites have started working. Eventually, nitrite levels will fall, and nitrates will begin to accumulate, indicating the third and final stage of the cycle is underway.
- **Wait for the Cycle to Complete:** The nitrogen cycle takes time—anywhere from 2 to 6 weeks, depending on your setup. Factors like tank size, water temperature, and the amount of ammonia added can influence how long it takes. You'll know the cycle is complete when both ammonia and nitrite levels drop to zero and you can detect a small amount of nitrate in the water. At this point, your tank is much safer for fish, as the bacteria colonies are established and can handle the bio-load.
- **Perform Water Changes if Needed:** During the cycling process, ammonia and nitrite levels may spike dangerously high. If this happens, perform small water changes (about 10-15%) to bring these levels down without completely resetting the cycle. However, don't overdo the water changes, as beneficial bacteria need time to grow. You're trying to strike a balance between allowing the bacteria to develop and keeping water conditions from becoming too toxic.
- **Wait for Nitrate Levels:** Once ammonia and nitrite levels reach zero, you should see nitrates building up. Nitrates are much less harmful, but it's important to keep them in check, as high nitrate levels can still stress fish and encourage unwanted algae growth. Regular water changes will help keep nitrate levels under control. Aim for nitrate levels below 40 ppm, ideally closer to 10-20 ppm for a healthy, balanced aquarium.
- **Introduce Fish Gradually:** After your tank has cycled, it's time to start adding fish. However, it's crucial to do this gradually. Start with just a few fish and monitor your water parameters closely for any signs of ammonia or nitrite spikes. By adding fish slowly, you give the beneficial bacteria a chance to adjust to the increased bio-load, which helps prevent sudden changes in water quality.

The Importance of Patience

One of the biggest challenges when starting the nitrogen cycle is the wait. It's tempting to rush the process, but patience is key to success. A rushed nitrogen cycle can lead to an unstable tank, where ammonia and nitrites spike, putting your fish and plants at risk. Many aquarists find it helpful to think of cycling as the foundation of their aquascape—a step that ensures the long-term health of their aquatic ecosystem.

During this period, you might feel like nothing is happening, but behind the scenes, an essential ecosystem is forming.

Beneficial bacteria are establishing colonies on the filter media, in the substrate, and on other surfaces within the tank. These bacteria will work tirelessly to maintain water quality once fish and plants are introduced.

The Bottom Line

Setting up an aquascape is a rewarding process, but it requires careful planning and attention to detail. From sketching your initial design to cycling the aquarium, each step plays a crucial role in creating a balanced, thriving ecosystem. By following the step-by-step guide outlined in this chapter, you'll be well on your way to building an aquascape that not only looks beautiful but also provides a healthy environment for your plants and fish.

As you move forward with your aquascaping journey, remember that patience and observation are essential. Each phase, from the initial setup to the final stages of cycling, requires time to ensure long-term success. The more care you invest in the early stages, the more enjoyable and sustainable your aquascape will be in the years to come. Happy aquascaping!

CHAPTER 07

AQUASCAPE **MAINTENANCE** AND **CARE** TIPS

"A balanced ecosystem should require minimal intervention. By allowing natural processes to occur, you can maintain an aquascape that is both beautiful and sustainable."

Diana Walstad

So, you've created a beautiful aquascape filled with vibrant plants and active fish, and it's sitting in your home like a living work of art. But the journey doesn't stop once the aquascape is set up. Just like any living ecosystem, an aquascape requires ongoing care and attention. Without proper maintenance, the crystal-clear water could turn murky, plants could lose their luster, and algae could take over, detracting from the beauty and balance you worked so hard to achieve.

In this chapter, we'll dive into the practical side of aquascaping — maintenance. Regular care routines are essential for keeping your aquascape healthy and looking its best. Whether you're a beginner or an experienced aquascaper, knowing how to manage water changes, prune plants, control algae, and monitor water parameters will ensure that your aquascape thrives over the long term. By following these tips, you'll be able to enjoy a vibrant, balanced aquarium that's both visually appealing and ecologically sound. Let's begin with understanding the essential routines that will keep your aquascape in top shape.

Daily, Weekly, and Monthly Maintenance Routines

Maintaining a vibrant and balanced aquascape requires regular care, just like tending to a garden. This section focuses on the crucial role that daily, weekly, and monthly maintenance routines play in keeping your freshwater aquascape healthy. These consistent routines not only help prevent problems but also ensure your aquarium remains a beautiful, thriving ecosystem. In this section, you'll learn how to break down your aquascape care into manageable tasks that you can seamlessly integrate into your life.

The Importance of Routine Maintenance

Think of your aquarium as a living organism, one that relies on balance and consistency to flourish. Without regular care, it's easy for small issues to snowball into bigger problems, such as algae overgrowth, poor water quality, or stunted plant growth. These problems can affect the overall health of your fish and plants, as well as the aesthetic appeal of your aquascape. Establishing a routine will make it easier for you to spot changes early and address them before they become serious.

Water Changes: How Often and How Much

One of the most critical aspects of maintaining a healthy aquascape is performing regular water changes. Water changes help to remove waste products, toxins, and excess nutrients that can accumulate in the tank over time. Without these regular shifts, harmful substances can build up, impacting the health of your fish and plants.

- **Daily:** For high-tech setups, especially those that use CO2 injection, checking the water level daily is crucial. Water evaporates, especially in warm environments or under intense lighting. By topping off the tank, you maintain the proper water volume and prevent the concentration of dissolved solids, which could harm your aquatic life.
- **Weekly:** For most aquariums, changing 20-30% of the water every week is ideal. Weekly water changes help refresh the tank while maintaining a stable environment for fish and plants. When doing so, use a siphon to remove water from the tank while vacuuming the substrate. This helps get rid of detritus and leftover food that may otherwise rot and raise the nutrient levels, encouraging unwanted algae growth.
- **Monthly:** In larger or low-tech aquariums, you may not need to perform water changes as frequently. A 30-40% water change once a month might suffice. However, it's essential to regularly test water parameters to ensure that ammonia, nitrate, and pH levels remain in check. Drastic water changes can shock your fish, so be cautious. Always try to match the new water's temperature and pH to the existing water to avoid stressing your aquatic life.

Water conditioners are helpful when preparing water for your aquarium, as they neutralize chlorine and chloramine in tap water. This step is crucial because untreated water can be harmful to both fish and plants. The key is consistency—making smaller water changes frequently helps to maintain balance rather than drastic, infrequent shifts.

Pruning Plants: A Key to Balance and Aesthetic

Pruning is more than just an aesthetic choice—it's an essential practice for maintaining plant health. Plants in aquascapes can grow rapidly, sometimes outpacing the tank's ability to sustain them. If left unchecked, fast-growing plants can overshadow slower-growing species or clog your aquascape, disrupting the visual balance and hindering light distribution.

- **Daily:** Take a few moments each day to inspect your plants. Remove any dead or decaying leaves you notice. Doing so prevents organic waste from breaking down in the tank, which could increase the nutrient load and encourage algae blooms.
- **Weekly:** Fast-growing species, like Rotala or Ludwigia, may require pruning every week or two, especially in tanks that utilize high levels of CO_2 and light. Using aquascaping scissors, trim these plants to maintain their size and shape. For carpeting plants, like Monte Carlo, give them a "haircut" by cutting the top layer. This promotes lateral growth, keeping them compact and low to the substrate. Be mindful not to uproot these delicate plants in the process.
- **Monthly:** Once a month, take the time to re-evaluate your aquascape's overall design. Plants grow and change, which may require you to thin out certain areas or reposition plants to maintain the original layout. You can also propagate healthy cuttings to fill in gaps, ensuring your aquascape remains visually cohesive and healthy.

Cleaning the Glass, Substrate, and Filter

A clean aquarium is not only visually appealing but also healthier for your aquatic inhabitants. Over time, algae and biofilm can accumulate on surfaces, reducing light penetration and degrading water quality. Regular cleaning helps keep your tank looking pristine while preventing the buildup of harmful substances.

- **Daily:** Use a soft sponge or an algae scraper to gently clean the inside of the glass. Doing this daily prevents the algae and biofilm from building up, which can become harder to clean if left for too long. Be especially careful with acrylic tanks, as they can scratch easily.
- **Weekly:** During your weekly water change, use a gravel vacuum to clean the substrate. This removes any detritus, plant trimmings, and uneaten food that may have settled on the tank floor. Detritus can contribute to poor water quality, encouraging algae growth and reducing oxygen levels. Be mindful of delicate plants and their roots to avoid uprooting them during this process.
- **Monthly:** Once a month, you should clean your filter to maintain its efficiency. Filters are the backbone of your aquarium's water filtration system, and cleaning them prevents blockages and keeps the water flowing smoothly. To clean the filter, rinse the sponges and filter media in old aquarium water, not tap water. The beneficial bacteria that live in the filter media are vital for breaking down waste,

and tap water's chlorine can kill them. Additionally, check for clogs in hoses or impellers and inspect filter parts for wear and tear.

Why Regular Maintenance is Worth the Effort

You might wonder if sticking to these daily, weekly, and monthly routines is necessary. The truth is that consistency in maintenance translates to fewer problems down the road. By regularly checking water parameters, trimming plants, and cleaning the tank, you reduce the risk of algae outbreaks, nutrient imbalances, or equipment failure. Furthermore, a well-maintained aquascape enhances the well-being of your fish, who thrive in clean, stable environments.

Not only does routine maintenance benefit your aquatic life, but it also allows you to enjoy your aquascape at its best. Clear water, vibrant plants, and healthy fish create a calming and visually stunning environment that can be a source of pride and relaxation.

Managing Algae

When it comes to maintaining a beautiful aquascape, algae often feels like the persistent villain. While some algae growth is natural and expected in any freshwater aquarium, excessive algae can quickly become an eyesore, competing with your plants for nutrients and ruining the overall look of your aquascape. Managing algae is essential to keeping your tank healthy and visually appealing. By understanding the types of algae you may encounter and how to control them, you can prevent your aquascape from becoming overrun.

Common Types of Algae and Control Methods

Different types of algae thrive under different conditions, so it's important to know what you're dealing with to take the right steps toward control.

- **Green spot algae:** If you notice small green dots on the glass or slow-growing plants, you're likely dealing with green spot algae. This type of algae is typically caused by too much light or an imbalance in phosphate levels. To manage it, try reducing the amount of light your aquarium receives and increasing CO2 levels. You might also benefit from using a phosphate buffer to keep levels in check. Regularly cleaning your tank's glass will also help prevent the algae from building up.
- **Hair algae:** This type of algae looks like long, wispy strands that cling to plants and hardscape materials. It's often a sign of nutrient imbalance, particularly an excess of iron or light. Start by manually removing as much hair algae as possible. Then, adjust your fertilization routine to avoid overloading the tank with iron. Additionally, adding algae-eating creatures like Siamese algae eaters or Amano shrimp can help keep this problem under control.
- **Black beard algae (BBA):** Black beard algae is particularly stubborn, often growing in thick, dark patches on plant edges and hardscapes. BBA is usually associated with low CO2 levels or poor water circulation. Increasing the amount of CO2 in your tank and improving water flow can help eliminate this pesky algae. Another effective treatment is spot-treating affected areas with liquid carbon, which can weaken the algae and make it easier to remove.

- **Brown algae (diatoms):** If your tank looks like it's been dusted with brown powder, especially in the early stages of setup, you're probably dealing with brown algae. This algae is common in new tanks and tends to disappear on its own as the aquarium matures. To help speed up the process, slightly increase the light exposure and use high-quality water to reduce silicate levels, which contribute to brown algae growth.

Preventing Algae Growth with Lighting and CO2 Management

The best way to combat algae is to prevent it from taking over in the first place. The key to this is maintaining the right balance between lighting, nutrients, and CO2. When one of these elements is out of balance, algae will often begin to grow uncontrollably. Here's how you can keep things balanced:

- **Limit lighting:** Too much light is one of the main culprits behind algae blooms. Aim to keep your aquarium lights on for about 8 hours a day, especially in high-tech setups. Avoid extending the lighting period, as this can give algae the perfect conditions to thrive. If you notice excessive algae growth, reducing the light duration or intensity can make a significant difference.
- **Consistent CO2 use:** For aquariums that use CO2 injection, it's crucial to ensure that CO2 levels remain stable and sufficient for your plants. Low CO2 levels create an opportunity for algae, like black beard algae, to take over. Regularly monitor CO2 levels and adjust as necessary to promote plant growth and inhibit algae.
- **Balance nutrients:** Over-fertilizing your tank can lead to nutrient imbalances, which encourage algae growth. Be mindful of your dosing, particularly with nutrients like iron and phosphate, which algae tend to thrive on. Follow recommended guidelines for fertilizers, and test your water regularly to ensure that levels remain optimal for plants, not algae.

Maintaining an Algae-Free Aquascape

Preventing algae requires a combination of routine maintenance and careful monitoring of your tank's environment. Regular water changes, pruning of plants, and checking water parameters will help you stay ahead of potential algae issues. By maintaining the right balance of light, CO2, and nutrients, you can enjoy a healthy, algae-free aquascape that looks stunning and provides a thriving environment for your aquatic life.

Water Parameters

Monitoring and maintaining proper water parameters is essential for the health of both plants and fish. Water quality can fluctuate over time due to factors like fish waste, plant decay, and added nutrients, so it's important to keep an eye on key water parameters.

Testing and Maintaining pH, KH, GH, and Nutrient Levels

Regularly testing your water ensures that the parameters remain within safe limits. Here's what you should be monitoring:

- **pH:** The pH of your water affects the availability of nutrients to your plants and the health of your fish. Most planted tanks do well with a pH between 6.5 and 7.5, but some plants and fish may have specific pH requirements. Use a pH test kit to check your water regularly and adjust as needed with pH buffers or other treatments.
- **KH** (Carbonate Hardness): KH refers to the level of carbonates and

bicarbonates in the water, which act as a buffer to prevent sudden pH changes. A KH level of 4-8 dKH is generally good for most aquascapes, as it provides stability for the pH. Low KH can lead to pH swings, which can stress plants and fish. Use KH test kits and buffers to maintain stable levels.

- **GH** (General Hardness): GH measures the concentration of calcium and magnesium in the water, both of which are important for plant and fish health. Most aquascapes thrive with a GH level of 3-10 dGH. Some species of plants and fish, however, may prefer softer or harder water. You can adjust GH by adding minerals or using specialized water conditioners.
- **Nutrient Levels:** In planted aquariums, it's essential to monitor key nutrients like nitrate, phosphate, and potassium to avoid deficiencies or excesses. Plants need these nutrients for growth, but imbalances can lead to algae problems or poor plant health. Test kits for nitrate and phosphate are available and should be used to keep nutrient levels in check. Aim for 10-20 ppm of nitrate and 0.5-1 ppm of phosphate in your tank, and adjust fertilization accordingly.

Feeding Fish and Plant Fertilization

In any well-maintained aquascape, finding the right balance between feeding your fish and fertilizing your plants is essential. Overfeeding or improper fertilization can lead to nutrient imbalances, causing algae growth and other issues. Conversely, underfeeding results in malnourished fish and plants, ultimately impacting the health and beauty of your aquatic environment. This section explores how to approach feeding and fertilization to ensure both your fish and plants thrive.

Types of Fish Food and Proper Feeding Practices

Feeding your fish seems simple, but getting it right is crucial for maintaining water quality and overall tank health. One of the most common mistakes aquascapers make is overfeeding. Uneaten food quickly sinks to the substrate, breaking down and releasing excess nutrients that contribute to algae blooms. To avoid this, it's important to feed your fish just enough to meet their nutritional needs without leaving waste behind.

- **Flake Food:** Flakes are a staple for most aquarium fish and are easy to portion. Since they float on the water's surface, they're ideal for mid- and top-feeding fish. However, if left uneaten, flakes can become soggy and settle at the bottom, so it's important to remove any leftovers after feeding.
- **Pellets:** If you have bottom feeders like Corydoras, pellets are a great option. Pellets sink more quickly than flakes, making them accessible to fish that prefer scavenging at the tank's base. Be mindful not to overfeed pellets, as they can decompose quickly if not eaten.
- **Frozen or Live Food:** Incorporating frozen or live foods like bloodworms or brine shrimp adds variety to your fish's diet, mimicking their natural feeding habits. These are excellent as occasional treats, providing high-quality nutrients and promoting more vibrant fish. However, because live food can introduce bacteria or parasites, make sure to source from reputable suppliers and avoid overfeeding.

A good rule of thumb is to feed your fish what they can consume within two to three minutes, once or twice a day. If you notice excess food collecting at the bottom of the tank, it's a sign that you're likely overfeeding. Always be prepared to remove uneaten food promptly to prevent it from degrading water quality and fueling algae growth.

Plant Fertilization: Supporting Healthy Growth

Plants in your aquascape require a careful balance of nutrients to grow and maintain their vibrant appearance. Just as overfeeding your fish can cause nutrient imbalances, improper fertilization of plants can lead to problems like algae overgrowth. There are several methods for delivering nutrients to plants, depending on their needs and the type of aquascape you maintain.

- **Liquid Fertilizers:** Liquid fertilizers are commonly used in aquariums with plants that draw nutrients from the water column. These fertilizers typically contain essential nutrients like nitrogen, phosphorus, and potassium. Following the manufacturer's dosing guidelines is crucial to avoid over-fertilizing, which can trigger algae blooms. In high-tech tanks with fast-growing plants, regular testing of water parameters can help you fine-tune the amount of fertilizer you're adding.
- **Root Tabs:** Some plants, like Amazon Swords or Cryptocorynes, absorb nutrients primarily through their roots. Root tabs are small fertilizer capsules placed directly into the substrate, slowly releasing nutrients to the root zone. This method is particularly effective for rooted plants that require a steady supply of nutrients but may not benefit as much from liquid fertilizers.
- **CO2 Injection:** If you're running a high-tech aquarium, CO2 injection becomes a critical part of your plant care routine. Plants use CO2 for photosynthesis, and in setups with intense lighting, CO2 helps promote faster and healthier plant growth. A CO2 concentration of 20–30 ppm (parts per million) is considered ideal for most planted tanks. However, CO2 levels need to be monitored carefully to ensure they remain stable, as fluctuations can stress both plants and fish.

Maintaining Balance in the Aquarium

Striking the right balance between feeding your fish and fertilizing your plants is key to keeping your aquascape healthy. Regular dosing with liquid fertilizers or root tabs, combined with proper CO2 levels, ensures your plants have everything they need to thrive. At the same time, being mindful of how much you feed your fish prevents excess nutrients from accumulating in the tank, reducing the risk of algae outbreaks.

Remember, consistency is essential. By sticking to a feeding and fertilization schedule that's appropriate for your tank, you'll create a stable environment where both your fish and plants can flourish. Regularly monitor your tank's water parameters, observe how your plants and fish respond to changes and adjust your routine as needed to maintain balance. This attention to detail will pay off, helping you maintain a beautiful, vibrant aquascape.

Troubleshooting Common Issues

Even with the best care, problems can arise in your aquascape. Recognizing early signs of trouble and knowing how to fix them will help you keep your tank healthy. Here are some of the most common issues you may encounter and how to fix them:

Yellowing Leaves

Yellowing leaves are often a sign of nutrient deficiency, particularly nitrogen or iron. Check your fertilizer dosing and adjust accordingly. If the problem persists, test your water for nitrate and iron levels to ensure they are within the appropriate range.

Stunted Growth

Stunted plant growth can occur due to a lack of CO_2, insufficient light, or nutrient deficiencies. If your plants aren't growing as expected, first check your light intensity and duration. Plants may not be receiving enough light, especially in low-tech setups. Also, ensure that your CO_2 levels are adequate and that you're dosing fertilizers regularly.

Fish Stress

Fish stress can be caused by poor water quality, overcrowding, or sudden changes in water parameters. Common signs of stress include erratic swimming, clamped fins, or loss of color. To reduce stress, ensure that water parameters (pH, temperature, and ammonia levels) are stable, and provide plenty of hiding spots with plants and hardscape. Perform regular water changes and avoid overcrowding the tank.

Melted Plants

"Melt" occurs when plants, especially new ones like Cryptocoryne, lose their leaves after being introduced to a new tank. This is usually due to a shock from changes in water parameters. If your plants are melting, give them time to adjust to the new environment. Ensure that water parameters are stable, and remove any dead leaves to prevent decay.

Root Rot

Root rot can result from plants being buried too deeply in the substrate or poor circulation around the roots. If you notice blackened, mushy roots, gently remove the affected plants and trim away the damaged roots. Replant them, ensuring the roots are not buried too deep, and improve water flow in the tank to prevent the issue from recurring.

Floating Plants Getting Waterlogged

Floating plants like Water Lettuce and Frogbit may become waterlogged, causing them to sink. This can happen if they are exposed to too much surface agitation or lack proper light. Adjust your water flow to minimize surface disturbance and ensure that the plants are getting enough light to thrive. Regularly remove excess debris that may weigh them down.

The Bottom Line

Maintaining an aquascape requires time, patience, and attention to detail. From daily tasks like wiping down the glass and feeding your fish to weekly water changes and plant pruning, regular care is essential for keeping your aquascape vibrant and healthy. Managing algae, maintaining proper water parameters, and addressing common issues before they become major problems are all part of the journey. By establishing a consistent maintenance routine and being attentive to the needs of your plants and fish, you'll create an aquascape that not only looks stunning but thrives for years to come. The beauty and tranquility that a well-maintained aquascape brings are well worth the effort. With the right knowledge and practices, you can fully enjoy the underwater world you've created.

CHAPTER 08

COMMON **CHALLENGES** AND **SOLUTIONS**

"One of the challenges in aquascaping is over-complicating things. Sometimes, allowing nature to take its course and keeping the design simple can resolve many issues."

Diana Walstad

Every aquascaper, whether a beginner or a seasoned hobbyist, eventually encounters challenges that can threaten the balance and beauty of their aquarium. From algae overgrowth to fish diseases, these problems can seem daunting, but they are all solvable with the right knowledge and approach. Addressing these challenges promptly and effectively is crucial not just for maintaining the aesthetics of your aquascape but also for ensuring the health and well-being of the living organisms within it.

This chapter dives deep into the most common challenges faced by aquascapers and offers practical solutions for overcoming them. We'll explore how to manage algae overgrowth, address plant health issues, tackle water quality problems, and prevent fish diseases. Finally, we'll talk about how to balance the delicate ecosystem of your tank, ensuring that your plants, fish, and bacteria coexist in harmony.

By the end of this chapter, you'll be equipped with the knowledge to not only solve common problems but also prevent them from happening in the first place. Let's dive into the most common issues and their solutions.

Dealing with Algae Overgrowth

Algae overgrowth is a challenge that nearly every aquascaper will face at some point. While algae naturally exist in any aquarium, when they grow out of control, they can quickly take over, affecting both the appearance and health of your aquascape. So, how do you deal with it? It all comes down to understanding what causes algae to grow and making the necessary adjustments to prevent it from taking over.

Identifying the Cause: Light, CO2, or Nutrient Imbalance

The first step in controlling algae is figuring out what's causing it. Algae thrive in environments where light, CO_2, and nutrient levels are out of balance. If you can pinpoint which factor is out of line, you'll be much better equipped to fix the problem.

- **Excessive Lighting:** One of the most common causes of algae overgrowth is excessive light. If you're new to aquascaping, it's easy to assume that more light equals better plant growth. However, too much light can backfire by encouraging algae to grow instead. You should aim to give your aquarium about 8–10 hours of light per day. If you're exceeding this, it's time to reduce the lighting period. You could also consider using a timer to ensure consistency. If your algae problem persists, try lowering the intensity of your light or switching to a bulb that has a spectrum more suitable for plant growth than algae.
- **CO2 Imbalance:** If the CO_2 in your tank is too low, your plants can't photosynthesize properly, allowing algae to outcompete them for nutrients. In high-tech setups where CO_2 is injected, you should aim for levels in the 20–30 ppm range. Maintaining this balance ensures that your plants get what they need while algae remain under control. If you don't use CO_2 injection, you can still help manage algae by increasing water flow and oxygenation, which supports plant growth and discourages algae.
- **Nutrient Imbalance:** Algae also thrive when there's too much or too little of certain nutrients. An excess of nitrates and phosphates often fuels algae growth. On the other hand, nutrient deficiencies weaken your plants, making them less effective at absorbing what's available, which leaves more for algae. Regularly test your water to monitor nutrient levels. If nutrients are too high, more frequent water changes can help reduce them. If nutrients are too low, you may need to adjust your dosing schedule or use a liquid fertilizer to boost plant health without encouraging algae.

Tailoring Your Solution to the Type of Algae

Not all algae are the same. Each type of algae has its own triggers, so it's essential to know what you're dealing with to effectively combat it. For instance, if you're seeing green spot algae on your plants or glass, it could indicate that phosphate levels are too low. On the other hand, black beard algae (BBA) is often a sign of low CO_2 levels. If BBA is popping up in your tank, increasing your CO_2 injection or improving water flow might be necessary.

Refer to Chapter 7 for a comprehensive guide on identifying and managing different algae types. By identifying the algae and understanding the conditions that promote its growth, you'll be much better equipped to handle the problem.

Long-Term Prevention

Once you've gotten your algae problem under control, the key is to prevent it from coming back. Consistency is crucial. Stick to a regular lighting schedule, maintain balanced CO_2 levels, and monitor nutrient levels closely. You should also make sure your plants are healthy and thriving, as robust plant growth naturally helps to keep algae in check by outcompeting it for resources. Healthy plants mean fewer opportunities for algae to take hold.

To sum it up, dealing with algae overgrowth in your aquascape requires careful observation and adjustments. By managing light, CO_2, and nutrients and identifying the type of algae you're dealing with, you can keep your tank balanced and algae-free. Remember, maintaining this balance is a long-term commitment, but once you get it right, your aquascape will thrive, and algae will no longer be a constant headache.

Plant Health Issues

Plants are the foundation of any aquascape, providing both aesthetic beauty and helping maintain balance in your freshwater aquarium. However, even the most carefully planned setups can face plant health issues, like yellowing leaves or stunted growth. Diagnosing these problems early and making the necessary adjustments is key to keeping your plants thriving.

Diagnosing Nutrient Deficiencies and Correcting Them

Plants, like any living organism, require the right balance of nutrients to grow properly. When certain nutrients are lacking, your plants will start showing signs of stress, such as discolored leaves or slow growth. Here are some of the most common nutrient deficiencies and ways to address them:

- **Nitrogen Deficiency:** Nitrogen is essential for plant growth. If the older leaves on your plants are turning yellow and dying off, it could be a sign of nitrogen deficiency. This is common in tanks with a low fish population since fish waste is a primary source of nitrogen. To correct this, you can add a nitrogen-based fertilizer or slightly increase your fish load to naturally introduce more nitrogen into the system.
- **Phosphorus Deficiency:** When plants lack phosphorus, you'll often notice dark green leaves with reddish or purple undertones. In more severe cases, older leaves may develop dead patches. If you suspect a phosphorus deficiency, you can use a phosphate-based fertilizer. However, too much phosphorus can encourage algae growth, so be mindful when dosing to avoid overdoing it.
- **Potassium Deficiency:** Potassium deficiency often causes small pinholes to form in older leaves, and you might notice yellowing around the leaf edges. Potassium isn't typically present in tap water, meaning you'll need to dose your aquarium regularly with

a potassium fertilizer to ensure your plants get what they need. Without adequate potassium, plant growth will slow, and overall plant health will decline.

- **Iron Deficiency:** Iron deficiency is one of the most common micronutrient issues in planted tanks. When iron is lacking, new leaves will often turn yellow or pale, while older leaves stay green. To fix this, you can add a liquid iron supplement. If you're using tap water that's low in minerals or soft water, you may need to supplement iron more frequently.

Regular Monitoring and Adjustments

Keeping a close eye on your plants and adjusting nutrient levels as needed is essential for maintaining a healthy aquascape. Testing your water regularly for nitrates, phosphates, and other key nutrients allows you to identify and address deficiencies early. By understanding and responding to the needs of your plants, you can ensure they get the proper balance of nutrients to thrive.

Regular water testing also helps prevent over-fertilization, which can lead to algae growth. By striking the right balance between light, CO2, and nutrients, your plants can outcompete algae for resources, creating a stable, healthy environment for both plants and fish.

Water Quality Problems

Water quality is the cornerstone of a thriving aquascape. When your water is balanced and clear, your plants grow more robustly, your fish stay healthier, and you minimize algae outbreaks. However, when things go wrong, water quality issues can quickly spiral into bigger problems, threatening the delicate balance of your tank. You might find yourself facing challenges like cloudy water, fluctuating pH levels, and hard water conditions. The good news is that these problems, though frustrating, are usually solvable with some attention to detail and the right approach.

Cloudy Water

Cloudy water is a common issue in both new and established tanks. It's not just an eyesore but can indicate underlying problems that need to be addressed. If your water becomes cloudy shortly after setting up a new tank, you're likely dealing with bacterial blooms. This happens as beneficial bacteria work to establish themselves, and it's a normal part of the tank cycling process. The key here is patience—this cloudiness usually clears up on its own as the tank stabilizes. To help the process along, make sure your filtration system is functioning properly, and avoid overfeeding your fish, as excess food can decompose and contribute to cloudiness.

On the other hand, if your tank is established and suddenly becomes cloudy, the culprit might be suspended fine particles, especially if you've recently disturbed the substrate. A quick fix for this is to use a fine filter pad that can capture these particles and help restore your water's clarity. Regular water changes and vacuuming the substrate can also help prevent particles from clouding your tank in the first place.

pH Swings

Maintaining a stable pH is critical for the health of both your fish and plants. Large swings in pH can cause stress and potentially lead to disease or death. pH fluctuations often occur when the carbonate hardness (KH) of your water is too low. KH acts as a buffer, preventing drastic pH changes. Without it, your water's pH can shift suddenly, especially after water changes or the addition of certain chemicals.

You can stabilize your pH by raising your tank's KH. One simple method is to add crushed coral to your filter or use it as part of your substrate. Crushed coral slowly releases calcium carbonate, which helps buffer the water and stabilize pH. Alternatively, commercial buffer solutions are available, allowing you to adjust KH with more precision. Ideally, for a freshwater aquascape, you want to aim for a stable pH range of 6.5 to 7.5, though this may vary depending on the specific needs of your fish and plants. Regular monitoring with a reliable pH test kit can help you catch any potential swings before they become harmful.

Hard Water

Hard water, characterized by high levels of calcium and magnesium, can be particularly challenging in aquascaping. While some fish and plants thrive in hard water, others struggle to grow or stay healthy. Additionally, hard water can lead to the buildup of calcium deposits on your equipment, making maintenance more difficult.

To manage hard water, many aquarists turn to reverse osmosis (RO) systems. These systems filter out excess minerals, softening the water and giving you more control over its hardness. If you don't want to install a full RO system, you can dilute your tank water with RO water purchased from a store. Just be sure to adjust the general hardness (GH) gradually, as sudden changes can shock your plants and fish.

Regular testing of your water's hardness is crucial. A GH test kit will allow you to track calcium and magnesium levels, and you can make small adjustments as needed to create the optimal environment for your aquascape.

Regular Maintenance is Key

Ultimately, keeping your water quality in check requires diligence and regular maintenance. Investing in a reliable water test kit is essential—it allows you to monitor parameters like ammonia, nitrate, pH, and hardness on a regular basis. By catching issues early, you can make adjustments before they escalate into serious problems that affect the health of your tank.

Incorporating small but consistent changes, like routine water changes and equipment checks, will ensure that your aquascape remains a vibrant, healthy environment for both plants and fish. Water quality might be a complex aspect of aquascaping, but with the right tools and knowledge, it's completely manageable.

Fish Health and Disease Prevention

Fish are the focal point of many aquascapes, and keeping them healthy should be a top priority. However, like any living creature, fish can become stressed or sick, especially in a poorly maintained environment. Recognizing the early signs of disease and knowing how to treat them can make all the difference in keeping your fish happy and healthy.

Identifying and Treating Common Fish Diseases

Fish diseases can develop quickly and spread through a tank if not addressed promptly. Here are some common fish diseases and how to treat them:

- **Ich (White Spot Disease):** Ich is one of the most common fish diseases caused by a parasitic protozoan. Fish with ich develop small white spots on their bodies and may scratch against surfaces. Treat ich by gradually raising the tank temperature to around 86°F (30°C) and using over-the-counter ich medications. Ensure proper water circulation and oxygenation during treatment.

- **Fin Rot:** Fin rot is a bacterial infection that causes the edges of a fish's fins to fray and disintegrate. Poor water quality or injury can lead to fin rot. Improving water conditions and using antibacterial treatments can help cure the infection. In severe cases, quarantine the affected fish and treat them separately.
- **Swim Bladder Disease:** Fish with swim bladder disease struggle to maintain their buoyancy, often floating to the surface or sinking to the bottom of the tank. Overfeeding or poor diet is usually the cause. Fasting the fish for a couple of days and then feeding them fiber-rich foods like peas can help alleviate symptoms.
- **Fungal Infections:** Fungal infections typically appear as white or cotton-like patches on the fish's skin. These infections can be treated with antifungal medications and by improving water quality to prevent recurrence.
- **Velvet Disease:** Velvet is caused by a parasite that gives fish a gold or dusty appearance on their skin. Affected fish may scratch against objects or exhibit lethargy. Treatment involves using copper-based medications and turning off the tank lights, as the parasite thrives in light.
- **Dropsy:** Dropsy is a condition where a fish's body swells due to internal fluid buildup, often caused by bacterial infections. Affected fish may display raised scales and bloating. Treatment usually involves isolating the fish and using antibacterial medications. Maintaining good water quality helps prevent dropsy from occurring.
- **Gill Flukes:** Gill flukes are parasitic worms that attach themselves to a fish's gills, causing difficulty in breathing. Infected fish may display rapid gill movement, gasping for air at the surface, or rubbing against surfaces. Treatment involves using antiparasitic medications such as praziquantel and improving water quality to reduce stress on the fish. Quarantine is recommended to prevent the spread to other tank inhabitants.
- **Anchor Worms:** Anchor worms are parasitic crustaceans that burrow into a fish's skin, causing irritation and visible, thread-like protrusions. Infected fish may scratch or rub against objects, and you may notice redness or inflammation around the worms. Manual removal of the worms with tweezers, followed by antiparasitic medication and improved water conditions, is the usual treatment.

Preventing disease in the first place is always the best approach. Regular water changes, proper filtration, and not overstocking your tank will help reduce stress on your fish and keep their immune systems strong.

Balancing the Ecosystem

Maintaining balance in your aquascape is about more than just keeping the water clean or feeding your fish. It's about ensuring that every element—plants, fish, bacteria, and water chemistry—works in harmony. Imbalances in CO2, oxygen levels, or water flow can cause problems that ripple through your entire ecosystem. Let's explore how to maintain this delicate balance.

Managing CO2 Levels, Water Flow, and Oxygen

In planted tanks, CO2 is essential for photosynthesis, helping plants grow and thrive. However, too much CO2 can lead to low oxygen levels, which can harm fish. Aim for CO_2 levels between 20–30 ppm (parts per million) and use a CO_2 drop checker to monitor this. If your fish are gasping at the surface, it may indicate that your CO_2 levels are too high, and you should either reduce the CO_2 input or increase water circulation to improve oxygenation.

Proper water flow is crucial for distributing nutrients and CO_2 throughout the tank and for preventing dead spots where debris can accumulate. However, too much flow can stress fish and uproot plants. Use adjustable filters or powerheads to ensure even water circulation without creating strong currents that disturb your aquascape. If you notice plants swaying too much or fish struggling against the current, reduce the flow.

Ensuring that your fish have enough oxygen is vital. Plants produce oxygen during the day through photosynthesis, but at night, they consume oxygen just like fish. If your tank is densely planted or if you're running a high-tech setup with CO_2 injection, it's essential to ensure proper aeration, especially during nighttime. You can achieve this by increasing surface agitation with a filter outflow or using an air pump to create more oxygen exchange. Maintaining a balanced ecosystem means regularly monitoring and adjusting these elements to meet the needs of both your plants and fish. By fine-tuning CO_2, water flow, and oxygen levels, you can create a stable environment where all aspects of the aquascape thrive together.

The Bottom Line

Aquascaping is a dynamic and ever-evolving art form, and with it comes a set of challenges that every aquascaper will face at some point. Whether it's battling algae overgrowth, diagnosing plant nutrient deficiencies, or dealing with fish diseases, understanding the root cause of these issues and taking proactive steps to address them is essential for maintaining a healthy and beautiful aquarium.

The key to overcoming these challenges is patience and observation. By monitoring your aquascape regularly, making adjustments as needed, and learning from each experience, you'll become more adept at solving problems before they spiral out of control. Remember, aquascaping is as much about maintaining balance as it is about creating beauty. When you can master both, you'll be rewarded with a thriving, harmonious ecosystem that will bring joy and tranquility to your home for years to come.

As you continue your aquascaping journey, don't be discouraged by setbacks. Each challenge you face is an opportunity to learn more about the intricate interplay between water, plants, fish, and the environment. With the right knowledge and a bit of persistence, you'll be well-equipped to handle whatever comes your way, ensuring that your aquascape remains a vibrant and healthy part of your living space.

CHAPTER 09

ADVANCED **AQUASCAPING** TECHNIQUES

"In advanced aquascaping, the use of slope is crucial to create depth and perspective. Raising the substrate towards the back of the tank enhances the sense of scale and dimension."

George Farmer

Aquascaping, at its core, is about creating stunning underwater landscapes that blend the beauty of nature with the art of design. Whether you're looking to create a lush, plant-filled scene or a minimalist, rock-dominated masterpiece, advanced aquascaping techniques can take your aquarium to the next level. For experienced hobbyists, mastering these techniques allows for more control over the aesthetic and functional aspects of an aquascape, giving you the power to create living works of art that thrive over time.

This chapter delves into some of the most advanced techniques in aquascaping, including the intricate design principles of Dutch aquascaping, the minimalist beauty of Iwagumi layouts, the authenticity of biotope aquascaping, and the technical mastery of CO2 management and lighting. Each of these approaches requires a combination of creativity, knowledge, and attention to detail. By the end of this chapter, you'll have a clear understanding of how to apply these techniques to your own aquascapes, whether you're aiming to recreate a natural habitat or design a precise, highly manicured layout. Let's explore the world of advanced aquascaping and uncover the techniques that can elevate your aquarium into a masterpiece.

Dutch Aquascaping: Creating a Plant-Focused Layout

When you first dive into aquascaping, the Dutch style can feel like an art form that sets itself apart. It's not about driftwood or rocks dominating the scene; instead, it's a plant lover's paradise. Dutch aquascaping originated in the Netherlands during the 1930s and focuses almost entirely on the arrangement of plants, creating lush, vibrant underwater gardens. The goal is to build a structured, garden-like display where carefully chosen aquatic plants are the stars of the show. Each plant is meticulously selected and placed, not just for its beauty but for its role in the overall composition. In this style, hardscape elements like rocks or driftwood are minimized, keeping the focus squarely on the plants.

A successful Dutch aquascape draws the eye with its bold color contrasts, distinct plant groupings, and intricate details. It's about making an underwater garden that feels both natural and deliberate, a balance between wild beauty and meticulous planning. But this style isn't just about aesthetics—it's also about patience, as creating and maintaining a Dutch aquascape takes time and dedication.

Choosing the Right Plants and Colors

One of the first steps to creating a Dutch aquascape is selecting a wide variety of plant species. Dutch aquascaping often incorporates 10 or more different plants, which allows for diverse textures, colors, and growth patterns. When choosing plants, you want to think about how they'll complement one another. Contrasting colors, varied textures, and thoughtful placement are the heart of this style.

Dutch aquascaping is known for its striking color contrasts. You'll often see deep reds,

bright greens, and soft yellows working together to create visual interest. For instance, a common combination might be the deep red of Rotala Rotundifolia paired with the bright green of Limnophila Sessiliflora. This contrast not only draws the eye but also creates defined sections that give your aquascape a structured, organized look. To make the colors pop, it's important not to place too many plants of similar hues next to each other. Think of it like painting—each color should stand out on its own while contributing to the overall picture.

Aside from color, the texture of plants plays a big role in creating visual interest. Leaf shapes can add depth and variety to your aquascape. Broad-leafed plants like Echinodorus provide a nice contrast to finer, more delicate species like

Myriophyllum. Mixing textures, whether smooth, spiky, or feathery, prevents the layout from becoming monotonous. It also helps create focal points that draw the eye to specific areas of your tank.
Height is another key consideration in Dutch aquascaping. Plants are often arranged in layers, with taller plants in the background and shorter ones in the foreground. This height variation helps make the aquascape feel more three-dimensional and ensures that each plant is visible. Taller plants like Vallisneria often serve as a backdrop, while smaller species like Cryptocoryne can be used in the foreground to create a balanced, layered effect.

Layering and Grouping for Depth

To create depth in a Dutch aquascape, layering plants from front to back is essential. A well-executed Dutch aquascape typically features three distinct layers: foreground, midground, and background. Each layer has a specific role in the overall composition. For the foreground, low-growing plants like Cryptocoryne or Hemianthus Callitrichoides (commonly known as "dwarf baby tears") work well. These plants form a carpet-like effect that covers the bottom of the tank, providing a base layer for the aquascape. Moving to the midground, you'll want medium-height plants like Alternanthera Reineckii or Hygrophila Corymbosa. These species serve as a transition between the low-growing plants in the foreground and the taller plants in the background. Finally, tall plants like Rotala or Vallisneria are placed at the back of the tank to frame the entire layout.

Grouping plants is another hallmark of Dutch aquascaping. Unlike some other aquascaping styles where plants are more randomly placed, Dutch aquascaping relies on organized rows or sections of plants. This not only creates structure but also allows each plant to shine. Grouping plants with similar colors or textures together helps to create a cohesive look while also making it easier to maintain the layout over time.

Keeping It Pristine: Maintenance

Maintaining a Dutch aquascape requires consistent care and attention to detail. Because this style is plant-heavy, regular pruning is necessary to keep the plants looking neat and healthy. Fast-growing stem plants, in particular, need to be trimmed weekly to prevent them from taking over the tank. Pruning also encourages bushier growth, which helps maintain the organized look of a Dutch aquascape.

Water changes are another important part of maintaining a Dutch aquascape. Since the plants in this style grow rapidly, they consume a lot of nutrients. Regular water changes help to replenish these nutrients and keep the water chemistry stable, which is essential for healthy plant growth. A weekly water change of about 30% is often recommended to keep the tank in top condition.

In addition to pruning and water changes, you'll need to keep an eye on plant health. Yellowing leaves, slow growth, or algae buildup can indicate nutrient deficiencies or other issues. Monitoring your plants and making adjustments as needed will help keep your Dutch aquascape thriving.

The Takeaway

Dutch aquascaping is all about precision and creativity. It offers a unique way to showcase the beauty of aquatic plants by creating a structured, colorful, and vibrant underwater garden. The attention to detail required for plant selection, placement, and maintenance makes it one of the most rewarding styles of aquascaping.

If you're someone who loves working with plants and enjoys the challenge of creating a balanced, organized layout, Dutch aquascaping could be the perfect style for you. With careful planning and regular maintenance, you can create an aquascape that is both beautiful and sustainable.

Iwagumi Aquascaping: The Minimalist Approach

If you're looking for an aquascaping style that embodies the "less is more" philosophy, Iwagumi aquascaping is the perfect example. While Dutch aquascaping is all about the vibrant use of plants and colors, Iwagumi is its minimalist counterpart, focusing on simplicity, balance, and the power of stone arrangements. Rooted in Japanese design principles, this style strips away the excess and finds beauty in restraint, allowing you to create serene underwater landscapes that evoke calm and tranquility. The word "Iwagumi" itself means "rock formation," which underscores the importance of rocks in this approach.

At the core of Iwagumi lies a deep connection to nature, where every stone placement feels intentional, creating a natural flow within the aquarium. As with many Japanese design concepts, it's not just about visual appeal—it's about balance, harmony, and the space between the elements. Achieving this serene beauty takes a careful eye for detail and a commitment to simplicity.

Choosing and Placing the Perfect Stones

In Iwagumi aquascaping, the rocks are the stars. Every stone contributes to the overall aesthetic, so thoughtfully selecting and placing them is essential. It's not just about picking rocks that look interesting; it's about how these stones interact with each other and with the plants and space around them. This is where your creativity comes into play.

Types of Rocks for Iwagumi

The types of rocks you use can have a dramatic effect on the feel of your aquascape. Seiryu stone, Dragon stone, and Ryuoh stone are among the most popular choices for Iwagumi layouts. Seiryu stone, with its deep ridges and blue-gray color, is especially favored because of its natural, sculpted look. These rocks add texture and create a sense of age and permanence, giving your aquarium an ancient, weathered feel. Dragon stone and Ryuoh stone are also great choices, offering different textures and colors that can set the tone for your design.

Stone Selection and Placement

When selecting your rocks, start by choosing a primary stone, often called the "Oyaishi." This stone is the largest and most eye-catching, serving as the focal point of your aquascape. The Oyaishi should be placed slightly off-center, as this creates a more natural and pleasing look. An off-center focal point allows for a sense of movement and flow in the design, drawing the viewer's eye naturally across the layout. Too much symmetry can make the aquascape feel static and artificial. Alongside the Oyaishi, you'll select several smaller stones, known as "Fukuishi," to complement the main stone. These supporting stones are placed around the Oyaishi in a way that enhances its size and shape. The key is to arrange them in a natural, asymmetrical pattern. Odd numbers of stones, like three or five, are typically used in Iwagumi because odd numbers tend to feel more balanced and dynamic to the human eye. By avoiding symmetry, you create a more organic, flowing design that feels like a snapshot of nature.

Substrate Sloping for Depth

In addition to the rocks, the substrate plays an important role in creating depth and perspective in an Iwagumi layout. By sloping the substrate from back to front, you enhance the visual impact of the rocks, making the layout feel more dynamic and three-dimensional. This subtle incline helps guide the viewer's eye naturally from the foreground to the background, enhancing the sense of flow and movement within the tank.

Achieving Balance in a Minimalist Design

At first glance, Iwagumi aquascaping might seem simple, but don't let that fool you. The minimalism of this style is deceiving—achieving balance and harmony with so few elements can be challenging. Every stone placement, every slope in the substrate, and every plant must work together to create a design that feels effortless yet intentional.

Plant Selection

While rocks are the centerpiece of Iwagumi aquascaping, plants play a supporting role. The choice of plants in an Iwagumi layout is minimal, often limited to just one or two species. This is a stark contrast to Dutch aquascaping, where multiple plant species create a riot of colors and textures. In Iwagumi, simplicity reigns. Carpeting plants like Glossostigma Elatinoides or Eleocharis Acicularis are popular choices. These plants form a lush, even carpet across the substrate, helping to soften the rocks and provide a natural backdrop without detracting from the stones themselves.

The use of plants in Iwagumi is about creating contrast and accentuating the stones rather than competing with them. The minimalist plant selection keeps the focus on the rocks while still adding life and softness to the layout.

Achieving Balance and Flow

The beauty of Iwagumi lies in its ability to balance simplicity with complexity. The rocks may be few, but the way they are arranged creates a sense of flow that invites the viewer's eye to move naturally through the tank. The rocks and plants work together to create a visual journey that starts in the foreground and flows through to the background, much like a landscape painting.

To achieve this balance, it's important to avoid placing the main stone in the center of the tank. A centered stone can create a static, stagnant look. Instead, place it slightly off-center, letting the smaller stones and plants guide the eye across the layout. The key is to make the design feel organic, as though the rocks and plants naturally fell into place over time.

The Takeaway

Iwagumi aquascaping is as much about patience and precision as it is about creativity. It's a style that challenges you to create balance and harmony with minimal elements, teaching you to find beauty in restraint. The result is a tranquil, peaceful underwater landscape that invites reflection and calm.

If you're someone who appreciates simplicity and the natural beauty of stone, Iwagumi aquascaping is the perfect style for you. It may require an eye for detail and a steady hand, but the result is a strikingly beautiful aquascape that feels both timeless and serene.

Biotope Aquascaping

Biotope aquascaping is a style that goes beyond just creating a beautiful underwater scene. It's about bringing a slice of a specific natural habitat into your home and replicating it as closely as possible. Unlike more aesthetic-driven styles like Dutch or Iwagumi aquascaping, biotope aquascaping prioritizes realism. The goal is to mimic the exact conditions of a particular region or ecosystem, from the plants and fish to the water chemistry and even the substrate. Whether it's the dark, tannin-rich waters of the Amazon, the soft, acidic streams of Southeast Asia, or the rocky landscapes of Africa's Rift Valley, biotope aquascaping allows you to recreate an entire environment in miniature.

This style of aquascaping is perfect for hobbyists who love research and detail. It requires a deep understanding of the specific habitat you're replicating, making it both challenging and rewarding. As you dive into biotope aquascaping, you'll find yourself not only learning about aquatic life but also about geography, ecology, and the subtle dynamics that make each ecosystem unique.

Recreating Natural Habitats

When you're creating a biotope aquarium, the first step is to decide which habitat you want to replicate. Each habitat comes with its own unique challenges, but the effort pays off in a tank that feels like a genuine piece of nature. Three of the most popular biotope themes are Amazonian, Southeast Asian, and African.

Amazonian Biotope

An Amazonian biotope is one of the most iconic and popular setups for biotope enthusiasts. The Amazon River basin is home to a wide variety of species, and its conditions are distinct. The water is typically soft and acidic, often stained brown by tannins released from decaying plant matter and driftwood. When setting up an Amazonian biotope, you'll want to use a sandy substrate, along with lots of driftwood and perhaps some rocks, to create shaded areas and hiding spots.

Plants like Echinodorus (Amazon Sword) and Vallisneria are common choices, as they thrive in the soft, acidic conditions of the Amazon. Floating plants like Pistia (Water Lettuce) or Salvinia can be added to create a canopy that filters the light, mimicking the shaded environment of the river. As for fish, species like Neon Tetras, Angelfish, and Corydoras are all native to this region and thrive in these water conditions. The combination of tannin-stained water, natural driftwood, and these species will give your tank the authentic feel of an Amazonian riverbed.

Southeast Asian Biotope

A Southeast Asian biotope focuses on the soft, acidic waters of this diverse region. The plants and fish from Southeast Asia are well-adapted to these conditions, and they can create a tranquil, lush aquascape. Popular plant choices include Java Fern, Cryptocoryne, and various types of moss, all of which are native to the streams and rivers of Southeast Asia. These plants are relatively easy to care for and thrive in low to moderate lighting, making them perfect for replicating the dappled light conditions of a Southeast Asian stream.

For fish, species like Rasboras, Gouramis, and Bettas are native to Southeast Asia and fit perfectly into this type of biotope. The hardscape typically consists of driftwood and smooth river stones, mimicking the natural environment of the region. This setup creates a serene, natural look that feels close to what you might find in the shallow streams and rivers of Southeast Asia.

African Biotope

African biotopes often focus on the lakes and rivers of the Rift Valley, particularly those that house species like African Cichlids. Unlike the plant-heavy setups of other biotopes, an African biotope often features more rocks and caves, as the lakes of this region have hard, alkaline water with minimal plant life. The hardscape is the focal point here, with large rocks providing shelter for the fish.

African Cichlids are known for their vibrant colors and territorial behavior, and they require plenty of hiding places and territories within the tank. To replicate the conditions of their natural habitat, it's important to maintain the right water chemistry—hard, alkaline water with a high pH. This biotope is less about plants and more about creating a rocky landscape that allows the fish to thrive in a setup that mimics their natural environment.

Selecting the Right Flora and Fauna

One of the most critical aspects of biotope aquascaping is selecting the right plants and fish. Everything in the tank should come from the same region or ecosystem to maintain authenticity. This can be more difficult than in other styles of aquascaping, as you'll need to match water conditions like pH, hardness, and temperature to what these species would experience in the wild.

Plant Selection

When choosing plants for your biotope, it's important to stick to species native to the region you're replicating. For an Amazonian biotope, plants like Echinodorus and Vallisneria thrive in soft, acidic water, while Southeast Asian biotopes often feature Java Fern and Cryptocoryne. African biotopes, on the other hand, may have little to no plant life, so they focus on rocky hardscapes instead.

Selecting native plants ensures that your biotope is as realistic as possible. Introducing non-native species can disrupt the balance of the ecosystem you're trying to replicate and may create unnecessary challenges in terms of water chemistry and tank maintenance.

Fish Selection

Just as important as plant selection is choosing the right fish for your biotope. Fish species should match the water conditions and ecosystem you're trying to create. For instance, South American fish like Neon Tetras and Angelfish require soft, acidic water, while African Cichlids prefer hard, alkaline conditions.

Compatibility is key, both in terms of water parameters and behavior. Make sure that the fish species you select can coexist peacefully and thrive in the same environment. This not only adds to the authenticity of your biotope but also ensures the long-term health and well-being of your fish.

The Takeaway

Creating a biotope aquarium is a challenging yet deeply rewarding experience. It requires more research and planning than other styles, but the result is a tank that feels like a real slice of nature. Whether you're replicating the dark waters of the Amazon or the rocky shores of an African lake, biotope aquascaping allows you to bring the beauty and diversity of the world's ecosystems into your home. The attention to detail and authenticity make this style particularly appealing to aquarists who enjoy research and want to create an environment that is not only beautiful but also true to life.

Using CO2 Injection Systems

For advanced aquascapers, using CO2 injection is a game-changer. CO2 is one of the most important nutrients for plant growth, and by injecting CO2 directly into your tank, you can create a lush, fast-growing aquascape that wouldn't be possible with ambient CO2 levels alone. That said, managing CO2 effectively requires careful attention to balance.

Advanced CO2 Management for Healthy Plants

Using a CO2 injection system requires careful management. Too little CO2 and your plants won't thrive; too much, and you risk suffocating your fish. Advanced CO2 systems allow for precise control over CO2 levels, ensuring that your plants have everything they need to grow while maintaining a safe environment for your fish.

- **CO2 Injection:** Most advanced aquascapers use pressurized CO2 systems to maintain consistent levels. A CO2 cylinder, regulator, and diffuser work together to inject controlled amounts of CO2 into the water. It's important to monitor CO2 levels closely using a drop checker or pH probe to ensure they stay within the ideal range (20–30 ppm for most plants).
- · Distribution: Even distribution of CO2 throughout the tank is essential. Poor circulation can lead to "dead zones," where plants don't receive enough CO2 and start to wither. Using a good filter and strategically placed circulation pumps will help ensure that CO2 is evenly dispersed throughout the aquarium.
- **Monitoring CO2 Levels:** Use a CO2 drop checker to monitor the levels of dissolved CO2 in your aquarium. Ideally, you want to aim for a concentration of around 20–30 ppm. If you notice your fish gasping at the surface, it's a sign that CO2 levels are too high, and you should reduce the flow.

Tips for Balancing Light and Nutrients with CO2

CO2 alone won't lead to healthy plants—there must be a balance between light, nutrients, and CO2 levels. Too much light without adequate CO2 can lead to algae growth, while too little light can stunt plant development.

- **Lighting:** High-light setups generally require more CO2 to support rapid plant growth. If you're using intense lighting, you'll need to increase CO2 and nutrient dosing to keep up with the plants' demands. Conversely, in a low-light setup, less CO2 and fewer nutrients are needed.
- **Fertilization:** Plants need a balanced

supply of macronutrients (nitrogen, phosphorus, potassium) and micronutrients (iron, manganese, etc.) to grow effectively. CO_2 injection accelerates plant growth, which means your plants will need more nutrients to keep up. Regularly dose with liquid fertilizers or root tabs to provide essential nutrients like nitrogen, phosphorus, and potassium. Also, weekly water changes help prevent nutrient buildup and maintain a healthy environment.

By balancing CO_2, light, and nutrients, you can create an environment where plants flourish and outcompete algae. Proper CO_2 management is essential for any high-tech aquascape, especially when working with demanding plant species.

Lighting and Photoperiod Optimization

Lighting is one of the most critical factors in any aquascape, especially when you're working with high-tech setups that include CO_2 injection. Proper lighting not only supports plant growth but also enhances the colors and overall appearance of your aquascape.

Controlling Light to Enhance Plant Growth and Coloration

Light intensity, spectrum, and duration all affect plant growth and coloration. Different plants have different light requirements, so understanding how to control your lighting is key to creating a thriving aquascape.

- **Light Spectrum:** Most aquatic plants require light in the blue and red spectrum to photosynthesize efficiently. A color temperature of 6,500K (Kelvin) is generally ideal for most aquariums, as it closely mimics natural daylight. This range provides the full spectrum of light needed for photosynthesis, helping plants grow and enhancing their natural colors. LED lights specifically designed for plant growth often provide this spectrum, promoting healthy growth and enhancing the natural colors of your plants.
- **Photoperiod:** The duration of light exposure is just as important as the intensity. Most aquariums do well with about 8–10 hours of light per day. Avoid leaving the lights on for too long, as this can lead to algae growth. A consistent light schedule, controlled by a timer, ensures that your plants get the light they need without promoting algae.

- **Adjusting Light for Coloration:** If you're looking to enhance the reds and purples in your plants, such as in species like Rotala or Ludwigia, consider using stronger lighting and CO_2 supplementation. These plants require high light and nutrient levels to achieve their full-color potential.

Optimizing your lighting setup will not only improve plant health but also enhance the overall visual appeal of your aquascape. Proper lighting can bring out vibrant colors, improve plant growth, and help create a balanced, algae-free environment.

The Bottom Line

Advanced aquascaping techniques open up a world of possibilities for creating stunning, nature-inspired aquariums. Whether you're drawn to the structured beauty of Dutch aquascaping, the minimalist elegance of Iwagumi, or the natural authenticity of a biotope, mastering these techniques allows you to push the boundaries of your creativity.
Balancing CO_2, light, and nutrients is

essential for achieving lush plant growth, and using high-tech tools like CO2 injection systems and advanced lighting will help you create the perfect environment for your aquascape. With practice, patience, and a deep understanding of these techniques, you can take your aquascaping to new heights, creating underwater landscapes that are not only beautiful but also thriving ecosystems.

As you continue your aquascaping journey, remember that patience and persistence are key. The most beautiful aquascapes take time to develop, and each challenge you overcome will bring you closer to creating the perfect underwater world.
Now, it's time to experiment, refine your skills, and let your creativity flow as you embark on the journey of advanced aquascaping. The possibilities are endless, and each new challenge you tackle will bring you closer to creating your dream aquascape.

CHAPTER 10

AQUASCAPING **INSPIRATIONS** AND **GALLERY**

"Aquascaping galleries provide a treasure trove of ideas. You can learn so much by studying the work of other aquascapers and interpreting their techniques in your own way."

Oliver Knott

Aquascaping is both an art and a science, where nature meets design to create breathtaking underwater landscapes. Whether you're new to the hobby or a seasoned aquascaper, finding inspiration from other designs can spark new ideas and push your creative boundaries. Seeing the works of others, especially those that have gained recognition in the aquascaping community, can help you understand what makes an aquascape truly stand out. In this chapter, we'll showcase some of the most inspiring aquascapes from around the world and explore why they work so well.

But inspiration is only the first step. As the aquascaping community continues to grow, so do the opportunities for collaboration, learning, and even competition. We'll also dive into how you can get involved in the global aquascaping community, participate in competitions, and refine your skills. Whether you're aiming to create a personal masterpiece or compete on an international stage, this chapter will guide you on how to harness your passion for aquascaping and take it to the next level. Let's start by looking at some stunning aquascapes that have captivated audiences worldwide.

Showcasing Beautiful Aquascapes

Aquascapes come in all shapes and styles, from minimalist rock-based layouts to dense, plant-heavy designs that mimic natural riverbeds or forests. What these designs have in common is their ability to transport viewers into a serene, underwater world. By analyzing some of the most famous and awe-inspiring aquascapes, you'll see how design elements like balance, contrast, and flow come together to create a cohesive and beautiful whole. Let's take a closer look at some inspiring aquascapes from around the world and analyze what makes them so captivating.

Inspiring Layouts from Around the World

Aquascaping is a global phenomenon, and different regions often have their own unique styles and approaches. Whether it's the intricate plant-focused designs of Dutch aquascapes or the minimalist elegance of Japanese Iwagumi layouts, the variety is endless. Let's take a tour of some of the most stunning aquascapes from across the globe.

The Nature Aquarium by Takashi Amano (Japan)

Takashi Amano, widely regarded as the father of modern aquascaping, introduced the world to the "Nature Aquarium" style. His works are inspired by the beauty of the natural world, often mimicking landscapes you'd find in forests, mountains, or rivers. Amano's aquascapes are known for their incredible sense of depth and natural balance, created by using stones, driftwood, and carefully arranged plants.

In one of his most famous creations, the "Ryoboku" layout, Amano uses large pieces of driftwood to mimic the structure of a submerged tree, with plants like Java fern and Anubias growing organically around the wood. The arrangement of stones and plants creates a natural flow that draws the viewer's eye from one point to another, giving the impression of a living, breathing ecosystem. Amano's work has inspired generations of aquascapers, and his principles of using natural elements to evoke the essence of nature remain foundational in aquascaping today.

The Dutch Style Aquascape (Netherlands)

Dutch aquascaping, one of the oldest forms of planted tank design, is distinct for its focus on plants rather than hardscape. A Dutch aquascape is often compared to a well-tended garden, with rows of vibrant plants arranged in a carefully curated, almost architectural layout. Dutch aquascapes use a wide variety of plant species, often showcasing contrasting colors and textures to create visual interest.

In one standout example, the aquascaper used a combination of bold red plants like Alternanthera Reineckii, bright green Rotala, and subtle background species like Hygrophila to create layers of color. The key to a successful Dutch aquascape is the meticulous placement of plants to form "streets" or "paths" of color that guide the viewer's gaze through the tank. These paths give the illusion of depth and space, making even a small tank feel larger than it is.

Iwagumi Aquascape by George Farmer (UK)

Iwagumi, a minimalist style that focuses on rock placement, is a challenging yet rewarding aquascaping technique. One of the masters of Iwagumi, George Farmer, has created numerous stunning aquascapes that highlight the simplicity and elegance of this style. In his design "Serenity Peaks," Farmer uses Seiryu stones to form a central mountain range, with low-growing plants like Monte Carlo carpeting the foreground.

The beauty of Iwagumi lies in its simplicity, but achieving balance in rock placement is no easy feat. Each stone is positioned with careful attention to proportion, scale, and texture. In "Serenity Peaks," Farmer's use of the rule of thirds ensures that the largest stone, known as the "Oyaishi," serves as the focal point, while smaller supporting stones, or "Fukuishi," enhance the overall composition without overpowering it. The result is a harmonious and tranquil aquascape that evokes a sense of natural beauty.

"Mountain Peaks" - Dave Chow (Hong Kong)

Dave Chow's "Mountain Peaks" is an Iwagumi-style aquascape that captures the essence of a mountain range emerging from a calm, misty valley. Using large Seiryu stones, Chow creates a sense of grandeur and isolation, with the stones acting as towering peaks. The careful placement of small ground-covering plants mimics a fog-covered plain, emphasizing the scale and isolation of the rocky peaks. The minimal use of plants in this aquascape lets the stones shine, highlighting the importance of rock placement in Iwagumi designs.

"Amazon Riverbed" - Oliver Knott (Germany)

Oliver Knott's "Amazon Riverbed" is a biotope aquascape that faithfully recreates a section of the Amazon River. The layout features authentic Amazonian plants like Echinodorus and floating water lettuce, paired with natural driftwood and fine sand to mimic the riverbed. Knott populates the aquarium with fish species native to the Amazon, including Cardinal Tetras and Apistogramma, creating an authentic slice of the Amazon's ecosystem. The attention to detail in replicating the natural habitat makes this aquascape not only visually appealing but also an educational experience.

"Windswept Plains" - James Findley (United Kingdom)

James Findley's "Windswept Plains" captures the feeling of an open, windy landscape with long grasses flowing in one direction. Findley uses tall, fine-leaved plants like Eleocharis to create the windswept look, while sparse rock formations suggest a barren yet beautiful plain. The use of negative space in this aquascape gives it a sense of vastness, even in a relatively small tank. This design is a perfect example of how simplicity and thoughtful use of space can evoke powerful emotions.

Analyzing Successful Designs: What Works and Why

Looking at these aquascapes, it's easy to be captivated by their beauty, but understanding what makes them successful requires a deeper analysis. Let's break down some of the common elements that contribute to their effectiveness:

Balance and Harmony

One of the first things that draw viewers into an aquascape is the sense of balance and harmony. This doesn't necessarily mean symmetry but rather a natural flow that makes the tank feel cohesive. A well-balanced aquascape has no elements that seem out of place or overpowering. In Takashi Amano's nature-style aquascapes, for example, the placement of rocks, plants, and driftwood creates a natural flow that mimics how these elements would exist in the wild.

Depth and Perspective

Creating the illusion of depth in a small tank can be challenging, but it's one of the most important factors in making an aquascape feel dynamic and engaging. By using techniques like substrate sloping, careful plant placement, and foreground-to-background layering, aquascapers can give their tanks a sense of three-dimensionality. In George Farmer's Iwagumi aquascapes, the placement of rocks and the use of carpeting plants help to guide the viewer's eye from the front of the tank to the back, enhancing the sense of depth.

Flow

Flow refers to how the viewer's eye moves across the aquascape. In "Mountain Peaks," the rocks are positioned in a way that naturally guides the eye from one peak to another. This intentional flow creates a sense of movement and depth, making the viewer feel as though they're traveling through the landscape.

Contrast and Color

Color contrast plays a major role in Dutch aquascapes, where different plant species are selected for their vibrant hues. By pairing bright green plants with deep red or purple species, aquascapers can create visual interest and highlight specific areas of the tank. Similarly, the use of light and dark rocks in Iwagumi aquascapes can add contrast and drama to the layout.

Proportion and Scale

Understanding the proportion and scale of the elements in your aquascape is crucial. In smaller tanks, large rocks or tall plants can easily overwhelm the space, while in larger tanks, tiny plants or small stones can get lost. Successful aquascapes maintain a sense of proportion, ensuring that all elements work together to create a unified whole.

Realism

In biotope aquascaping, realism is key. Oliver Knott's recreation of the Amazon River is successful because it feels authentic—everything from the plant choice to the fish species contributes to the overall realism. This level of detail immerses the viewer in the habitat and makes the aquascape more than just a decorative tank; it's a living representation of nature.

Aquascaping Competitions and Community

One of the most exciting aspects of aquascaping is the vibrant global community that surrounds it. From local clubs to international competitions, aquascapers have countless opportunities to connect, share ideas, and showcase their work. If you're looking to take your aquascaping hobby to the next level, getting involved in the community is a great place to start.

How to Get Involved in the Aquascaping Community

Getting involved in the aquascaping community can be both fun and rewarding. Whether you're seeking advice or inspiration or just want to share your work with others, there are plenty of ways to connect with fellow aquascapers.

Join Online Forums and Social Media Groups

One of the easiest ways to connect with the aquascaping community is through online forums and social media platforms. Websites like Aquatic Plant Central, UKAPS (U.K. Aquatic Plant Society), and The Planted Tank offer active forums where members can ask questions, share their progress, and offer advice. Similarly, Facebook and Instagram have vibrant aquascaping communities where hobbyists from around the world post photos, exchange tips, and celebrate each other's work.

Attend Local Aquascaping Meetups

Many cities have local aquascaping clubs or aquarium societies where members can meet in person to discuss their setups, trade plants and fish, and participate in workshops. These meetups are a great way to get hands-on experience and learn from more seasoned aquascapers. Check with local aquarium stores or search online for groups in your area.

Participate in Aquascaping Workshops and Events

Aquascaping workshops are often hosted by local aquarium shops or international aquascaping experts. These workshops provide a great opportunity to learn new techniques, get feedback on your designs, and even build an aquascape under the guidance of a professional. Workshops vary in format, from one-day events to longer courses, and are available both online and in person. Events like Aquatic Experience in the U.S. or Aquaflora workshops in Europe feature professional aquascapers who share their knowledge and provide live demonstrations on techniques like rock placement, planting, and CO2 management.

Entering Competitions: Tips for Success

Aquascaping competitions offer an exciting opportunity to showcase your skills and gain recognition within the community. Competitions like the International Aquatic Plants Layout Contest (IAPLC) and Aquatic Gardeners Association (AGA) International Aquascaping Contest attract entries from aquascapers around the world, and competing in these events can help you push your limits and refine your craft.

Understand the Competition Rules and Criteria

Before entering any competition, it's crucial to read and understand the rules. Each competition has its own set of criteria for judging, including layout, plant health, fish selection, and overall impact. Make sure you know what the judges are looking for and plan your design accordingly.

Focus on Originality

While it's tempting to emulate the works of famous aquascapers, originality is often rewarded in competitions. Judges appreciate designs that bring something new to the table, whether it's a unique layout, an unusual combination of plants, or an innovative use of hardscape materials. Don't be afraid to take risks and put your own spin on traditional styles.

Pay Attention to Detail

In competitive aquascaping, details matter. From the placement of individual plants to the cleanliness of the glass, every aspect of your aquascape will be scrutinized by the judges. Make sure your tank is in top condition when submitting photos, and consider using professional photography to capture the full beauty of your design.

Experiment with Lighting and Photography

Presentation is everything in competitions. Once your aquascape is complete, experiment with different lighting setups to highlight the best features of your design. Proper lighting can enhance the colors of your plants and create shadows that add depth to the scene. Additionally, take high-quality photographs that accurately represent your aquascape. Use a camera with a macro lens to capture close-up details and ensure the overall composition of the photo is well-lit and free of distractions.

Learn from Feedback

Even if you don't place in the top rankings, participating in a competition is a valuable learning experience. Many competitions provide feedback from judges, which can help you improve your technique and understand what works and what doesn't. Use this feedback to refine your skills and come back even stronger in future competitions.

The Bottom Line

Aquascaping is a constantly evolving art form driven by creativity, passion, and a deep connection to nature. Whether you're drawing inspiration from the stunning works of Takashi Amano or exploring the minimalist beauty of Iwagumi, there's always something new to learn and discover. By immersing yourself in the aquascaping community and participating in competitions, you can push the boundaries of what's possible and share your work with others who share your passion.

As you continue your aquascaping journey, remember that every tank you create is a reflection of your unique vision. Take inspiration from the world around you, experiment with different techniques, and most importantly, enjoy the process. The possibilities in aquascaping are endless, and with the right inspiration and support, you can create an underwater world that is truly your own.

BONUS EXTRA!!!

Scan the QR CODE and GET Your Bonuses:

1.
The Ultimate Betta Fish Guide

2.
7 Mistakes Even Expert Aquascapers Make (and How to Avoid Them)

3.
Daily, Weekly, and Monthly Aquascape Maintenance Checklists

As SPAM Filters Are Pretty Crazy These Days...
...WHITELIST this Email Address!

aquascaping@floraandwaterpublishing.com

In This Way, Your Bonus Will Appear in the Main Folder of Your INBOX and They Will Not Be Buried Along With Other Advertisements in Your PROMOTION/SPAM Folder.

<u>HERE IS HOW TO DO IT:</u>

From Android Smartphone/Tablet
Open the Contacts App;
In the lower right corner, tap + (Add);
Enter the name and email address and then tap Save.

From iPhone/iPad
Open the Contacts App;
In the upper right corner, tap + (Add);
Enter the name and email address and then tap Finish.

CONCLUSION

As you reach the final chapter of The Aquascaping Master Collection: Freshwater Aquariums, take a moment to reflect on the journey we've taken together. From understanding the basics of aquascaping to diving deep into advanced techniques, we've covered a lot of ground. You've learned how to turn a blank aquarium into a thriving, beautiful underwater world. But more importantly, you've learned how to balance art with science, patience with creativity, and nature with design.

We began by exploring what aquascaping truly is—a fusion of artistry and nature. It's more than just arranging plants and rocks; it's about creating a living, breathing ecosystem that's both beautiful and self-sustaining. Whether you're interested in the minimalist Iwagumi style or the vibrant layers of a Dutch aquascape, the essence remains the same. Every aquascape is a reflection of its creator's vision, and the techniques you've learned allow you to bring that vision to life.

We discussed the fundamental equipment needed to start aquascaping, from choosing the right aquarium size and lighting systems to selecting the best filtration methods and CO2 setups. Each component plays a vital role in maintaining the delicate balance of your aquascape. With the right tools and knowledge, even the most complex ecosystems become manageable.

You've also delved into plant and fish selection, understanding how these elements interact with each other. From low-tech aquariums that thrive with hardy plants to high-tech setups that demand precise care, you now have the knowledge to select the perfect plants and fish to suit your desired style. You've also learned about the different styles of aquascaping and how to use design principles like the rule of thirds to create stunning visual effects.

Maintenance is another critical aspect we covered in detail. Aquascaping isn't just about setting up your aquarium; it's about nurturing it over time. Regular pruning, water changes, and nutrient management are all part of the journey. With these maintenance routines in place, you can ensure that your aquascape continues to thrive for years to come.

But aquascaping isn't without its challenges, and we didn't shy away from addressing them. Whether it's battling algae overgrowth, managing water quality, or troubleshooting plant health issues, you now have a toolkit of solutions to handle these common problems.

Finally, we ventured into advanced techniques that push the boundaries of aquascaping. From mastering the art of Dutch aquascapes to creating biotopes that mimic natural habitats, you've learned how to elevate your designs to the next level. These advanced methods aren't just for the experts—they're for anyone willing to experiment and explore new possibilities.

When we started this journey, you might have been unsure about where to begin. Aquascaping can feel overwhelming at first—there's so much to learn, from equipment and plant care to design principles and ecosystem balance. But as you progressed through each chapter, you've seen how all these elements fit together to create something truly special.

Think back to the introduction, where we talked about the beauty and therapeutic benefits of aquascaping. Now that you've gained hands-on experience and a deeper understanding of the craft, you can appreciate how aquascaping is both an art form and a source of relaxation. It's not just about building an aquarium; it's about building a space that brings joy and peace to your life.

By following the steps outlined in this book, you've not only developed a new skill but also a new way of seeing the world. Aquascaping teaches you patience, creativity, and the importance of maintaining balance—not just in your aquarium but in life itself.

Now that you've completed the book, it's time to put your knowledge into action. Whether you're setting up your very first aquascape or refining an existing one, take the lessons you've learned and start creating. Begin by sketching out a design, gathering your materials, and planning your layout. Don't rush the process—enjoy every step. Aquascaping is as much about the journey as it is about the final result.

For beginners, I encourage you to start with a simple setup. Choose easy-to-care-for plants like Anubias or Java Fern, and go with a low-tech system that doesn't require CO2 injection. Focus on getting the basics right—lighting, filtration, and water parameters. As you gain more confidence, you can experiment with more complex layouts and high-tech equipment.

If you're more experienced, challenge yourself with a new style or technique. Try your hand at an Iwagumi layout, or experiment with a biotope that mimics a natural ecosystem. Push the boundaries of what you think is possible, and continue to learn from each aquascape you create.

Remember, aquascaping is a hobby where there's always room to grow. Every tank you design teaches you something new, whether it's about plant care, fish behavior, or design principles. The more you practice, the more skilled you'll become—and the more beautiful your aquascapes will be.

As you move forward on your aquascaping journey, keep in mind that perfection isn't the goal—growth is. Aquascaping is an evolving art, and each aquarium you create will reflect your current skills, knowledge, and creativity. Some tanks will turn out exactly as you imagined, while others might take unexpected turns. Embrace those surprises. They often lead to the most rewarding results.

I hope this book has inspired you to see aquascaping not just as a hobby but as a way to connect with nature and express your creativity. The world of freshwater aquariums is vast, and there's always something new to discover. Whether you're experimenting with new plants, trying out different design techniques, or learning more about fish care, the possibilities are endless.

As you continue to explore aquascaping, don't be afraid to share your creations with others. Join online forums, participate in aquascaping competitions, or simply invite friends and family to admire your work. Aquascaping is a community, and there's so much to learn from fellow enthusiasts.

Thank you for joining me on this journey through the world of aquascaping. Your time, attention, and dedication mean everything. Writing this book was a labor of love, and knowing that you've taken the time to read it from cover to cover fills me with gratitude. I hope that you've found the information helpful, the advice practical, and the inspiration abundant.

Remember, aquascaping is an ongoing adventure. There will always be new challenges to face, new techniques to master, and new designs to bring to life. But with the tools and knowledge you've gained from this book, you're well-equipped to tackle them all.
So, what are you waiting for? It's time to roll up your sleeves, grab your aquascaping tools, and start creating the underwater masterpiece you've been dreaming of. The world of aquascaping is yours to explore—so go ahead and make it your own.
Happy aquascaping!

APPENDIX

Glossary of Key Terms

1. **Acrylic Tank:** A lightweight, durable aquarium made of acrylic, offering clearer viewing than glass. It's less prone to breakage but more easily scratched and is often used for custom-shaped aquariums.
2. **Algae Bloom:** A rapid increase in algae population in an aquarium, often caused by excessive light, nutrients, or poor water quality, leading to reduced oxygen levels and harming fish or plants.
3. **Alternanthera Reineckii:** A popular red aquatic plant known for its striking color contrast, typically used in high-light aquascapes. It requires CO2 supplementation and regular trimming to maintain its vivid appearance.
4. **Ammonia:** A toxic nitrogen compound primarily produced by fish waste and decaying matter in aquariums. It's harmful to fish, and its buildup can be controlled through filtration and regular water changes.
5. **Ammonia Spike:** A sudden increase in ammonia levels within an aquarium, often caused by overfeeding, overstocking, or poor filtration. It can be lethal to fish and requires immediate corrective action.
6. **Anubi**as: A hardy, slow-growing aquatic plant that thrives in low light conditions. It's commonly used in aquascaping, often attached to driftwood or rocks, and requires minimal maintenance.
7. **Aquarium Heater:** A device used to maintain a consistent water temperature in an aquarium, essential for the health of tropical fish. It typically includes a thermostat for precise temperature control.
8. **Aquarium Tank:** A glass or acrylic container designed to hold water, fish, plants, and other aquatic life. It serves as the primary structure for creating and maintaining an aquascape or aquarium ecosystem.
9. **Aquascape Design Principles:** Guidelines for creating visually balanced and natural-looking underwater landscapes. These include the use of focal points, the rule of thirds, plant layering, and negative space to enhance aesthetic appeal.
10. **Aquascape Maintenance:** Routine tasks required to maintain the health and beauty of an aquascape, including water changes, plant trimming, algae control, and monitoring water parameters like pH, ammonia, and nitrate levels.
11. **Aquascaping**: The art of arranging plants, rocks, and driftwood in an aquarium to create a visually appealing underwater landscape. It combines principles of gardening, design, and ecology for aesthetic and environmental balance.
12. **Aquascaping Competitions:** Events where aquascapers showcase their designs, judged on creativity, technical skill, and ecological balance. Popular competitions include the International Aquatic Plants Layout Contest (IAPLC) and the Aquatic Gardeners Association Contest.
13. **Aquascaping Resources:** Tools, books, online forums, and websites that provide information and guidance on aquascaping techniques, plant care, and aquarium maintenance. They are essential for both beginners and experienced hobbyists.

14. **Aquascaping Tools:** Specialized instruments like tweezers, scissors, and gravel vacuums are used for planting, trimming, and maintaining an aquascape. These tools help ensure precision and ease when crafting a balanced aquarium layout.
15. **Background Plants:** Taller plants are placed at the back of the aquarium to provide structure, depth, and a backdrop for midground and foreground plants. Common examples include Vallisneria and Hygrophila species.
16. **Bacterial Bloom:** A rapid growth of beneficial or harmful bacteria in an aquarium, usually occurring after water changes or overfeeding. It can make the water cloudy but often subsides as the ecosystem balances.
17. **Beneficial Bacteria:** Microorganisms essential for breaking down toxic waste (ammonia and nitrites) in an aquarium. These bacteria aid in the nitrogen cycle, converting harmful compounds into nitrates that plants can absorb.
18. **Betta Fish:** A popular freshwater fish known for its vibrant colors and flowing fins. Bettas are solitary, territorial creatures often kept in small tanks and prefer calm, warm water environments.
19. **Biotope Aqua**scape: An aquascaping style that replicates a specific natural habitat, including native plants, fish, and substrates. It's designed to mimic a real-world ecosystem, emphasizing ecological accuracy over aesthetics.
20. **Carbonate Hardness (K.H.):** A measure of the carbonate and bicarbonate levels in water, which affect its buffering capacity. Higher K.H. stabilizes pH levels, while lower K.H. can result in pH fluctuations, affecting aquatic life.
21. **CO2 Drop Checker:** A device used to measure carbon dioxide (CO2) levels in an aquarium. It provides a visual indication, changing color to show whether CO2 levels are optimal, too high, or too low for plant growth.
22. **CO2 Injection**: The process of adding carbon dioxide to an aquarium, usually through a pressurized system, to boost plant growth. It's commonly used in high-tech aquascapes to support demanding plants.
23. **Community Tanks:** Aquariums designed to house multiple fish species that coexist peacefully. Success depends on selecting fish with similar water parameter needs and compatible temperaments, ensuring a harmonious environment.
24. **Cryptocoryne:** A diverse genus of aquatic plants popular in aquascaping for their variety of colors and sizes. These slow-growing plants thrive in low light and are often used in midground areas of an aquascape.
25. **Depth and Perspective:** Techniques used in aquascaping to create the illusion of space and distance within an aquarium. Common methods include layering plants, using sloped substrates, and placing smaller plants in the foreground.
26. **Driftwood:** Natural wood often used in aquascaping to provide structure and visual interest. It also serves as an attachment point for plants like mosses and helps create a natural, organic look in an aquarium.
27. **Dwarf Gourami:** A small, brightly colored freshwater fish known for its peaceful temperament. Often kept in community tanks, dwarf gouramis thrive in planted aquariums with slow-moving water and plenty of hiding spots.
28. **Dwarf Hairgrass**: A popular carpeting plant used in aquascaping to create a lush, grassy foreground. It requires high light and CO2 for optimal growth and is known for its ability to spread across the substrate.
29. **Eleocharis:** A genus of grass-like plants, commonly known as hairgrass, often used in aquascaping for carpeting. It thrives in high-light conditions with CO2 supplementation

and creates a natural, flowing appearance in tanks.

30. **Emergent Aquascapes:** Aquascapes featuring plants that grow both submerged and partially above the waterline. This style mimics natural environments like riverbanks and creates dynamic visual interest with mixed aquatic and terrestrial elements.
31. **Fertilization:** The process of adding nutrients, such as nitrogen, phosphorus, and potassium, to an aquarium to support plant growth. Fertilizers can be in liquid form, root tabs, or through nutrient-rich substrates.
32. **Filtration** System: A critical component in aquariums that removes waste, toxins, and debris from water. It includes mechanical, biological, and chemical filtration to maintain water quality and support a healthy ecosystem.
33. **Fish Behavior:** The actions and patterns exhibited by fish in response to their environment, including feeding, schooling, territorial displays, and breeding. Understanding fish behavior helps maintain a peaceful, balanced aquarium.
34. **Fish Breeding:** The process of encouraging fish to reproduce in captivity, often requiring specific water conditions and breeding environments. Many species have unique breeding rituals, from egg-laying to livebearing.
35. **Fish Compatibility:** The ability of different fish species to coexist in the same aquarium without aggressive behavior. It depends on factors like size, temperament, diet, and water parameter preferences.
36. **Fish Population Control:** The management of fish numbers in an aquarium to avoid overcrowding, which can lead to stress, poor water quality, and health problems. Regular monitoring and responsible breeding practices help maintain balance.
37. **Fish Waste:** Organic material produced by fish, including uneaten food and excrement. It contributes to ammonia buildup in aquariums, requiring filtration and water changes to prevent toxicity.
38. **Floating Plants:** Aquatic plants that float on the surface of the water, such as duckweed or water lettuce. They provide shade, reduce algae growth, and offer shelter for fish but require occasional trimming.
39. **Fluorescent Lighting:** A type of aquarium lighting commonly used before the advent of LED lights. It provides adequate light for plant growth but generates more heat and consumes more energy than modern LED systems.
40. **Foreground Plants:** Low-growing plants placed at the front of an aquascape to create a carpet-like effect. Examples include Dwarf Hairgrass and Monte Carlo, which are used to enhance depth and visual appeal.
41. **Glass Tank:** A type of aquarium made from glass, known for its clarity, scratch resistance, and durability. It's heavier than acrylic but remains the most popular choice for hobbyists due to its reliability.
42. **Golden Ratio:** A design principle used in aquascaping that involves placing key elements in proportions that align with natural patterns, creating harmony and balance. It's commonly applied in layout design for focal points.
43. **Gravel Cleaner:** A tool used to siphon debris and waste from the substrate in an aquarium. It helps maintain water quality by removing detritus and preventing the buildup of harmful compounds.
44. **Green Plants:** Aquatic plants that contribute to the visual and ecological balance of an aquarium. They photosynthesize, producing oxygen and absorbing nutrients, helping control algae and support healthy fish environments.
45. **Hard Water:** Water with high mineral content, particularly calcium and magnesium. In

aquariums, hard water affects pH stability and is preferred by certain fish species, like cichlids, while others thrive in soft water.

46. **Hardscape:** Non-living elements like rocks, driftwood, and gravel are used in aquascaping to create structure and form. It serves as the foundation of the design, complementing plants and providing shelter for fish.
47. **High-Light Setu**p: An aquarium with intense lighting is necessary for growing demanding plants that require more light to thrive. High-light setups often require CO2 supplementation and regular fertilization to prevent algae growth.
48. H**igh-Tech Aquascap**ing: Aquascaping that involves advanced equipment like CO2 injection systems, high-intensity lighting, and automated fertilization. It supports fast-growing plants and complex designs but requires more maintenance and knowledge.
49. **Hygrophila:** A genus of fast-growing aquatic plants often used in the midground or background areas of aquascapes. Hygrophila thrives in high-light conditions and provides dense foliage that adds texture and volume to aquariums.
50. **Iwagumi Layout:** A minimalist aquascaping style focusing on rock formations as the primary design element. Typically, an Iwagumi layout uses three main stones and minimal plants to create a Zen-like, balanced composition.
51. **Java Fern:** A hardy, low-light aquatic plant commonly used in aquascaping. Java Fern attaches to rocks or driftwood and requires minimal care, making it a popular choice for beginners and low-tech tanks.
52. **Java Moss:** A versatile moss used in aquascaping to create naturalistic textures and fill spaces. It thrives in various light conditions, can attach to hardscapes, and is often used in breeding tanks as a shelter for fry.
53. **LED Lighting:** Energy-efficient, customizable aquarium lighting, widely used for its ability to support plant growth while enhancing the visual appeal of the tank. It offers a full spectrum of light and produces less heat than older systems.
54. **LED vs Fluorescent:** A comparison between modern LED lights, which are energy-efficient and customizable, and traditional fluorescent lighting, which is less efficient and generates more heat. LED systems are now favored in most aquascapes.
55. **Light Spectrum:** The range of colors emitted by aquarium lights affects plant growth and appearance. Full-spectrum lights mimic natural sunlight, promoting photosynthesis, while specific wavelengths, like red and blue, enhance growth and color vibrancy.
56. **Lighting Systems:** Essential tools in aquascaping, providing the necessary light for photosynthesis and plant growth. Systems include LED, fluorescent, and high-intensity setups, each offering different levels of control and customization.
57. **Liquid Fertilizer:** A nutrient solution added to aquarium water to promote healthy plant growth. It typically contains macro and micronutrients like nitrogen, potassium, and iron, which are crucial for aquatic plants, especially in high-tech setups.
58. **Livestock Selection:** The process of choosing fish, shrimp, or other aquatic creatures that complement an aquascape. It considers factors like compatibility, water parameters, how the species interact with the plants, and overall design.
59. **Low-Light Setup:** An aquarium with minimal lighting, suitable for growing low-maintenance plants like Anubias or Java Fern. Low-light setups require less energy and maintenance but limit the types of plants that can thrive.
60. **Low-Tech Aquascaping:** A simpler approach to aquascaping that avoids advanced equipment like CO2 systems and high-intensity lighting. It's ideal for beginners, focusing on hardy plants that require minimal care and maintenance.

61. **Macro Nutrients:** Essential nutrients required by aquatic plants in larger quantities, including nitrogen (N), phosphorus (P), and potassium (K). These nutrients are critical for plant growth, promoting healthy leaves, roots, and overall vitality in aquascapes.
62. **Micro Nutrients:** Trace elements such as iron (Fe), manganese (Mn), and zinc (Zn) that plants need in smaller amounts. Micronutrients support vital plant functions like chlorophyll production and enzyme activation, ensuring balanced growth in aquascapes.
63. **Midground Plants:** Plants placed in the middle section of an aquascape, between the foreground and background. They add depth and visual transitions, often featuring medium-sized species like Cryptocoryne and Hygrophila.
64. **Monte Carlo:** A popular carpeting plant in aquascaping known for its small, bright green leaves. It thrives in high-light, CO2-injected setups and spreads across the substrate to create a dense, lush carpet in the foreground.
65. **Nano Aquariums:** Small aquariums, typically under 10 gallons, designed for compact spaces. Nano tanks require careful monitoring due to limited water volume but can showcase intricate aquascapes with small fish and plants.
66. **Neon Fish Schooling:** The behavior of neon tetras and similar fish, where they swim together in coordinated groups. Schooling behavior provides protection from predators and creates a mesmerizing visual effect in aquariums.
67. **Neon Tetras:** Small, colorful freshwater fish known for their iridescent blue and red stripes. Neon tetras are peaceful, social fish, often kept in schools in community tanks, adding vibrant color to aquascapes.
68. **Nitrate Levels**: A measurement of nitrate concentration in an aquarium, a byproduct of the nitrogen cycle. High nitrate levels can stress fish and lead to algae blooms, so regular water changes help keep levels in check.
69. **Nitrogen Cycle:** The process through which beneficial bacteria convert toxic ammonia into nitrites and then into less harmful nitrates. The nitrogen cycle is essential for maintaining a healthy aquarium environment for fish and plants.
70. **Paludariums:** Hybrid aquariums that combine aquatic and terrestrial elements, often featuring both water and land-dwelling plants and animals. Paludariums create dynamic ecosystems that mimic riverbanks or rainforests, blending aquascaping with terrarium design.
71. **Partial Water Changes:** The regular practice of replacing a portion of the aquarium's water to remove toxins like ammonia and nitrates. This helps maintain water quality, supporting plant growth and fish health.
72. **pH Balance:** The measure of acidity or alkalinity in an aquarium. Maintaining the correct pH balance is crucial for the health of both plants and fish, with different species having varying pH preferences.
73. **pH Probe:** A tool used to measure the pH level of water in an aquarium. It provides real-time readings and helps aquascapers maintain stable pH conditions, ensuring a healthy environment for aquatic life.
74. **pH Swing:** Rapid changes in the aquarium's pH levels, often caused by water changes, CO2 fluctuations, or imbalanced K.H. pH swings can stress fish and plants, so maintaining stable water parameters is essential.
75. **Photoperiod:** The duration of time that an aquarium's lights are on each day. A balanced photoperiod, typically 8-10 hours, promotes healthy plant growth and reduces the risk of algae overgrowth in aquascapes.
76. **Plant Propagation:** The process of growing new plants from cuttings, division, or seeds.

In aquascaping, propagating plants helps expand greenery, maintain plant health, and allows for the creation of multiple aquascapes from one plant source.

77. **Plant Substrate:** A nutrient-rich material placed at the bottom of the aquarium to support plant root growth. Popular substrates include ADA Aqua Soil and Fluval Stratum, which provide essential nutrients to aquatic plants.
78. **Pygmy Corydoras:** Small, peaceful bottom-dwelling fish that thrive in community tanks. Pygmy Corydoras are popular in nano aquariums, helping keep the substrate clean while their active behavior adds life to the tank.
79. **Red Plants:** Aquatic plants with red or purple leaves, such as Alternanthera Reineckii or Ludwigia, are used to create striking contrasts in aquascapes. These plants often require high light and CO2 to maintain their vibrant color.
80. **Ripariums:** Aquariums that mimic the natural habitats found along riverbanks or wetlands, incorporating both submerged and emergent plants. Ripariums offer a unique blend of aquascaping and gardening, emphasizing lush greenery above and below the waterline.
81. **Rock Formation:** A key element in aquascaping, where rocks are strategically placed to create natural-looking landscapes. Rock formations provide structure, depth, and shelter for fish, often serving as focal points in aquascapes.
82. **Root Fertilizer:** Nutrient-rich pellets or tabs inserted into the substrate to provide essential nutrients directly to plant roots. Root fertilizers are crucial for heavy root-feeding plants, ensuring healthy growth in planted aquariums.
83. **Rotala Rotundifolia:** A popular stem plant known for its vibrant pink or red coloration. It thrives in high-tech aquariums with intense lighting and CO2, adding bold color to midground or background areas.
84. **Rule of Thirds**: A design principle used in aquascaping to create balanced, visually appealing layouts. The aquarium is divided into thirds, and key elements like rocks or plants are placed at intersections to guide the viewer's eye.
85. **Sand Substrate:** Fine-grained substrate used in aquariums to create a natural, soft appearance. It's often used in biotope aquascapes or with certain fish species that prefer sandy environments, like corydoras or loaches.
86. **Shrimp Tanks:** Aquariums specifically designed to keep freshwater shrimp species like cherry shrimp or Amano shrimp. Shrimp tanks often include dense vegetation and low-flow filters, providing optimal conditions for shrimp breeding and survival.
87. **Softscape:** The living components of an aquascape, including plants, mosses, and algae. Softscape contrasts with hardscape (rocks and wood) and plays a key role in creating a balanced and natural-looking aquarium environment.
88. **Species-Only Tanks:** Aquariums designed to house just one species of fish or invertebrates. This setup allows aquarists to focus on the specific needs and behaviors of the chosen species without the complications of managing multiple species.
89. **Staurogyne Repens:** A compact, green plant used in midground or foreground areas of aquascapes. It thrives in high-light, CO2-injected setups, forming dense bushes that add texture and depth to layouts.
90. **Substrate:** The material at the bottom of an aquarium, essential for plant growth and aesthetic purposes. It can be inert (like gravel or sand) or nutrient-rich, providing vital nutrients for rooted plants.
91. **Substrate Fertilizer:** Tablets or pellets are placed in the aquarium substrate to deliver nutrients directly to plant roots. These fertilizers help root-feeding plants thrive and are

commonly used in planted aquariums with nutrient-poor substrates.

92. **Temperature Control:** The process of maintaining stable water temperatures in an aquarium, typically using heaters or chillers. Temperature control is crucial for the health of both fish and plants, as sudden fluctuations can cause stress or disease.
93. **Tetras:** A popular group of small, colorful freshwater fish known for their schooling behavior. Tetras are peaceful, easy to care for, and often used in community tanks, adding vibrant color and activity to aquascapes.
94. **Trimming Tools:** Specialized scissors and tweezers used for pruning and maintaining aquatic plants. Trimming tools allow aquascapers to shape plants, control growth, and keep the aquarium layout neat and aesthetically pleasing.
95. **Underwater Ecosystem:** A balanced aquarium environment where fish, plants, bacteria, and other organisms coexist in harmony. Healthy underwater ecosystems require proper water parameters, nutrient cycling, and maintenance to thrive.
96. **Vallisneria:** A fast-growing, grass-like aquatic plant often used in the background of aquascapes. Vallisneria thrives in a variety of conditions and spreads quickly via runners, adding natural movement and texture to aquariums.
97. **Water Clarity:** The visual clearness of water in an aquarium is affected by factors like filtration, debris, and algae. Maintaining water clarity is important for both aesthetics and the health of fish and plants.
98. **Water Filtration**: The process of removing waste, toxins, and impurities from aquarium water using mechanical, biological, or chemical methods. Proper filtration is essential for maintaining a healthy, balanced environment for fish and plants.
99. **Water Hardness:** The concentration of dissolved minerals, primarily calcium and magnesium, in aquarium water. Water hardness affects pH stability and the health of certain fish species, with preferences varying from soft to hard water environments.
100. **Water Parameters:** These are key water quality metrics, including pH, ammonia, nitrites, nitrates, temperature, and hardness, that must be monitored and controlled to ensure the health of an aquarium's inhabitants.

Resources for Aquascaping Supplies and Plants

Entering Competitions: Tips for Success

Aquascaping is more than just a hobby; it's an art form. Whether you're a beginner looking to start your first planted tank or an experienced aquarist ready to take your designs to the next level, finding the right resources is key. Having access to high-quality supplies and the healthiest plants can make all the difference in transforming a simple aquarium into a thriving underwater ecosystem.

But where do you start? The world of aquascaping can be overwhelming, with endless options for plants, substrates, lighting, and more. This guide is designed to help you navigate that world, offering you a curated list of the best resources for aquascaping supplies and plants. From online shops to trusted local nurseries, you'll discover where to find everything you need to bring your aquascaping vision to life. Let's dive in and explore the best places to get aquascaping supplies, plants, and the essential tools to craft your aquatic masterpiece.

Why Quality Supplies and Plants Matter

Before jumping into where to find supplies, let's address why quality matters. Aquascaping isn't just about creating something beautiful—it's about building a sustainable environment for aquatic life. The plants you choose must thrive, the substrate must support root growth, and the lighting must provide the energy your underwater plants need to grow. Cheap, low-quality materials can lead to poor plant health, algae blooms, or even the failure of the entire ecosystem.

When you invest in top-quality supplies, you're giving your plants and fish the best chance to thrive. Plus, working with high-quality materials makes the aquascaping process more enjoyable and less stressful, as you'll spend less time troubleshooting and more time enjoying your beautiful creation.

Where to Buy Aquascaping Supplies

1. Online Stores for Aquascaping Supplies

In today's digital world, online shopping offers the convenience of accessing a wide variety of aquascaping products from the comfort of your home. These online stores specialize in everything from tanks and lighting systems to substrates and fertilizers.

In Europe:

- **The Green Machine (U.K.):** This UK-based store is well-known in the aquascaping community for its selection of high-quality plants, substrates, and aquascaping tools.
- **Aquaessentials (U.K.):** A well-stocked UK-based online store offering everything from plants to CO2 systems and aquascaping tools.
- **Aqua-Shop (Germany):** Catering to German-speaking countries, Aqua-Shop specializes in ADA products and live aquatic plants.
- **Scaped Nature (Netherlands):** This Dutch store provides high-quality hardscape materials, plants, and aquascaping tools.
- **Aquasabi (Germany):** Aquasabi is a trusted source in Europe for aquascaping tools, high-end substrates, and rare aquatic plants.
- **Green Aqua (Hungary):** With a broad range of plants, lights, and aquascaping tools,

Green Aqua is a popular online store for European enthusiasts.

- **ProShrimp (U.K.):** Known for both shrimp and aquascaping supplies, ProShrimp provides a variety of tools, substrates, and additives.
- **Flowgrow (Germany):** Flowgrow offers an extensive collection of plants, fertilizers, and aquascaping accessories to hobbyists in Europe.
- **Aquarium Gardens (U.K.):** This UK-based retailer stocks plants, substrates, hardscape materials, and filtration systems ideal for aquascapes.
- **ADA France:** Specializing in ADA products, ADA France offers a wide range of substrates, fertilizers, and CO2 systems for aquascaping.
- **ScapeTime (Poland):** A Polish online store offering a variety of plants, hardscape materials, and aquascaping tools.
- **Aqua-Fleur (Netherlands):** Focused on aquatic plants, Aqua-Fleur provides a wide range of plant species and aquascaping tools.
- **Aquascaping Shop (Italy):** This Italian store specializes in aquascaping equipment, plants, and high-quality hardscape materials.
- **Tropica (Denmark):** A renowned provider of aquatic plants and fertilizers, Tropica also offers aquascaping tools and guidance.
- **Swiss Aquascape (Switzerland):** For hobbyists in Switzerland, Swiss Aquascape offers an array of plants, substrates, and tools.
- **Fisch & Reptil (Germany):** This online store has an excellent selection of aquascaping products, from substrates to lighting.
- **Aquaristikshop (Germany):** Offering everything from live plants to technical equipment, this store serves German aquascapers well.
- **Nature Aquascaping (U.K.):** A UK-based retailer offering an array of substrates, hardscapes, and aquascaping kits for all levels.
- **Aquaplants (Belgium):** A Belgian store offering aquatic plants, substrates, and all the necessary tools for aquascaping.
- **Planted Aquarium (Spain):** This Spanish store specializes in aquatic plants and tools specifically for aquascaping setups.
- **WaterScaping Shop (Sweden):** A Sweden-based online retailer focused on providing high-quality plants, substrates, and aquascaping tools.

In North America:

- **Buce Plant (U.S.):** Buce Plant is a highly recommended resource for aquascaping enthusiasts. They offer a broad range of live plants as well as essential aquascaping tools, substrates, and equipment you need to create your ideal aquatic environment.
- **Aqua Forest Aquarium (U.S.):** If you're looking for a specialized aquascaping retailer that caters to both beginners and professionals, Aqua Forest Aquarium in the U.S. is an excellent option.
- **Glass Aqua (U.S.):** Glass Aqua specializes in rare aquatic plants, hardscape materials, and aquascaping tools.
- **Aquarium Co-Op (U.S.):** A popular online store offering a variety of aquarium supplies, including aquascaping essentials.
- **Boost Plant (U.S.):** Boost Plant provides a wide range of live aquatic plants, substrates, and CO2 equipment.
- **Flip Aquatics (U.S.):** Known for shrimp but also offering a comprehensive range of aquascaping tools, substrates, and plants.
- **The Shrimp Farm (U.S.):** In addition to shrimp, this store offers an extensive selection

of aquascaping plants and hardscape materials.

- **Aquarium Plants Factory (U.S.):** Specializing in aquatic plants and aquascaping kits, Aquarium Plants Factory offers a large selection.
- **Aquatic Arts (U.S.):** Aquatic Arts provides live plants, CO2 systems, and everything needed to start or maintain an aquascape.
- **Green Leaf Aquariums (U.S.):** Green Leaf Aquariums focuses on CO2 systems, lighting, and specialized aquascaping equipment.
- **Aquatic Design Centre (Canada):** Located in Canada, this store offers a variety of aquascaping plants, tools, and technical equipment.
- **Bonsai Driftwood (U.S.):** This store is specialized in providing bonsai-style driftwood, a popular choice in aquascaping.
- **Modern Aquarium (U.S.):** Modern Aquarium offers a wide selection of live plants, substrates, and aquascaping tools.
- **PetSmart (U.S. and Canada):** This major pet retailer provides a variety of aquarium plants, hardscape materials, and equipment for beginners.
- **Aquarium Depot (Canada):** A Canadian online store specializing in aquatic plants and aquascaping tools.
- **The Planted Tank (U.S.):** The Planted Tank offers everything from substrate to lighting systems specifically for planted tanks.
- **AquaZone (Canada):** A Canadian store that focuses on providing aquascaping plants, lighting, and CO2 systems.
- **Big Al's (Canada and U.S.):** Big Al's provides a range of aquascaping tools, plants, and accessories across North America.
- **Aquarium Design Group (U.S.):** This US-based store is known for offering high-end aquascaping equipment and premium aquatic plants.
- **Aquatic Supplies Canada:** This store provides a comprehensive selection of plants, CO2 systems, and aquascaping materials.
- **Dallas Discus (U.S.):** In addition to fish, Dallas Discus offers plants, driftwood, and other materials ideal for aquascaping setups.
- **Carolina Aquatics (U.S.):** Carolina Aquatics provides a wide range of live plants and substrates, specializing in aquascaping tools.

2. Local Fish Stores (LFS)

While online shopping is convenient, there's something to be said about visiting your local fish store (LFS). Many LFS specialize in aquascaping supplies, and they often have knowledgeable staff who can provide valuable advice tailored to your local water conditions and climate.

- **The Value of Hands-On Experience:** Shopping locally allows you to see plants and materials in person before purchasing, which is crucial when it comes to assessing the health of live plants. You can inspect the condition of leaves, check for any algae or pests, and ensure you're bringing healthy specimens into your tank. Additionally, LFS often stock hard-to-find plants that may not be available online.
- **Supporting Local Businesses:** When you shop at your local fish store, you're supporting small businesses within your community. Many LFS are family-owned and operated, providing personalized service that you can't always find online. Plus, these stores are often passionate about aquascaping and can offer valuable advice and local tips.

Best Places to Buy Aquatic Plants

Choosing the right plants is one of the most exciting parts of aquascaping. There are countless plant species to choose from, each offering unique textures, colors, and growth patterns. Below are some of the best resources for buying healthy, high-quality aquatic plants.

Europe:

1. **Tropica (Denmark):** A leading European supplier of aquatic plants known for tissue culture plants and extensive plant selection.
2. **AquaEssentials (U.K.):** Offers a wide variety of aquatic plants and supplies and is known for its high-quality products.
3. **Green Aqua (Hungary):** Provides an array of aquatic plants, aquascaping tools, and ADA products.
4. **Aquasabi (Germany):** Popular for aquatic plants, hardscape materials, and aquascaping tools.
5. **The Green Machine (U.K.):** UK-based store specializing in aquatic plants and aquascaping materials.
6. **ProShrimp (U.K.):** Offers a range of aquatic plants along with shrimp and other aquascaping essentials.
7. **Flowgrow (Germany):** Known for its vast collection of aquatic plants and advice for aquascaping enthusiasts.
8. **Aqua-Fleur (Netherlands):** Specializes in aquatic plants, offering a wide variety of species for aquascaping.
9. **Dennerle (Germany):** A trusted name in Europe that provides aquatic plants and nutrient-rich substrates.
10. **Aquaristikshop (Germany):** A comprehensive store offering aquatic plants, substrates, and equipment for aquascaping.
11. **Swiss Aquascape (Switzerland):** A Swiss-based store offering a range of aquatic plants for aquascaping enthusiasts.
12. **PlantedBox (U.K.):** Offers a wide variety of live plants for aquascapes, including low-tech and advanced plant species.
13. **Aquaplants Online (Netherlands):** Provides high-quality plants and delivers across Europe.
14. **AquaShop (Germany):** A German store known for its vast collection of aquatic plants and ADA products.
15. **Scaped Nature (Netherlands):** Specializes in plants, hardscape materials, and aquascaping tools.
16. **AquaByNature (U.K.):** A UK-based store providing a wide range of live plants for aquascapes.
17. **Fisch & Reptil (Germany):** Offers aquatic plants and aquascaping supplies with a focus on quality and variety.
18. **AquaplantShop (France):** A French store specializing in aquatic plants and aquascaping accessories.
19. **WaterScaping Shop (Sweden):** A Swedish store offering aquatic plants, substrates, and aquascaping materials.

20. **ScapeTime (Poland):** A reliable source for aquatic plants, providing a variety of plant species and tools for aquascapers.

North America:

1. **Buce Plant (U.S.):** A popular online store specializing in aquatic plants, particularly rare and unique species.
2. **Aquarium Co-Op (U.S.):** Well-known for its selection of aquatic plants and aquascaping tools, catering to beginners and experts.
3. **Aquatic Arts (U.S.):** Provides a wide range of aquatic plants, including tissue culture and live plants.
4. **Boost Plant (U.S.):** Offers a large variety of aquatic plants, focusing on quality and health.
5. **The Wet Spot Tropical Fish (U.S.):** A store specializing in tropical fish and a great selection of live plants for aquascaping.
6. **Aquarium Plants Factory (U.S.):** Known for a vast selection of plants, including rare and exotic species for aquascaping.
7. **Modern Aquarium (U.S.):** Offers a wide selection of plants, from beginner-friendly species to advanced plants.
8. **Aquatic Design Centre (Canada):** Located in Canada, this store offers a wide range of aquatic plants and tools for aquascaping.
9. **Big Al's (Canada & U.S.):** A well-known chain with both online and physical stores offering aquatic plants and supplies.
10. **Green Leaf Aquariums (U.S.):** Specializes in aquatic plants and CO2 systems, making it a one-stop shop for aquascapers.
11. **Glass Aqua (U.S.):** Provides high-quality aquatic plants and tools for hobbyists focused on unique aquascaping designs.
12. **Dallas Discus (U.S.):** Primarily a fish retailer, it also offers a selection of aquatic plants for aquascapes.
13. **PetSmart (U.S. & Canada):** A large chain that carries a variety of aquatic plants, available both online and in-store.
14. **The Shrimp Farm (U.S.):** Known for its shrimp, it also offers a solid variety of live aquatic plants.
15. **Aquazone (Canada):** A Canadian store offering aquatic plants and aquascaping supplies for hobbyists.
16. **Liquid Gardens (U.S.):** Specializes in high-quality aquatic plants and offers products for aquascaping enthusiasts.
17. **Flip Aquatics (U.S.):** Known for shrimp, Flip Aquatics also provides a wide variety of aquatic plants.
18. **Carolina Aquatics (U.S.):** A reliable source for aquatic plants, offering both easy and advanced species.
19. **The Planted Tank (U.S.):** An online retailer with a focus on aquatic plants and related aquascaping tools.
20. **Aquarium Depot (Canada):** A Canadian store offering live aquatic plants and aquascaping equipment.

Key Aquascaping Supplies You Need

Once you've decided where to buy your aquascaping materials, the next step is knowing what to buy. Here's a breakdown of the essential supplies you'll need for a successful aquascape.

Aquarium and Stand

1. **Aquarium tank:**
    - o **Size:** The size of your tank is crucial, and the bigger, the better for aquascaping. A larger tank allows more space for plants, hardscape elements, and fish to thrive. Beginners often start with a 20-gallon tank, while experienced aquascapers may go up to 100 gallons or more.
    - o **Shape:** Rectangular tanks are most common for aquascaping as they provide a larger surface area for plant growth and more viewing angles. Cubic tanks can also be used for smaller setups.
    - o **Material:** Glass is the most common and scratch-resistant, while acrylic is lighter but more prone to scratching.

2. **Aquarium stand:**
    - o The stand must be able to support the tank's weight when filled with water, substrate, and decorations. Custom-built stands can include space for filtration systems and CO_2 canisters, allowing for a cleaner aesthetic.

Filtration System

3. **External or internal filter:**
    - o **Canister filters:** Offer powerful filtration for large tanks, providing superior mechanical, biological, and chemical filtration. They're external and keep the aquascape visually cleaner.
    - o **Hang-on-back (HOB) filters:** Convenient for smaller tanks, offering easy maintenance.
    - o **Sponge filters:** Ideal for planted tanks with shrimp or small fish, as they don't create too much water flow and provide gentle filtration.

4. **Filter media:**
    - o **Mechanical:** Foam pads to capture debris.
    - o **Biological:** Ceramic rings or bio-balls that house beneficial bacteria to break down waste.
    - o **Chemical:** Activated carbon to remove toxins, discoloration, and odors from the water.

Lighting

1. **Aquarium lighting:**
    - o **LED lights:** The most energy-efficient and widely used. Full-spectrum LEDs support photosynthesis and plant growth, promoting lush greenery. Adjustable lighting intensity is important for balancing plant health and algae prevention.
    - o **T5 fluorescent lights:** Another good option for planted tanks, but they tend to consume more energy.
2. **Timer for lights:**

o Automated timers ensure a consistent light cycle, typically 8-10 hours per day, preventing algae overgrowth and maintaining a healthy environment for plants.

Substrate

1. Aquascaping substrate:
   - o Aquasoil: Nutrient-rich and specifically designed for plant growth. It helps maintain water pH levels suitable for aquatic plants.
   - o Sand or gravel: Common for aesthetic layers but typically need root tabs or liquid fertilizers to provide plants with essential nutrients. Fine gravel is often used for detailed foreground work, while coarse gravel can be layered with nutrient-rich substrates.

2. Root tabs:
   - o These slow-release nutrient capsules are buried in the substrate to feed root-heavy plants like Amazon swords and crypts.

CO_2 System (Optional but Beneficial for Heavily Planted Tanks)

1. **CO_2 injection system:**
   - o CO_2 supplementation promotes lush plant growth by increasing the availability of carbon, a key element in photosynthesis. It's essential for densely planted aquascapes or if you're keeping demanding species such as carpeting plants.

2. **CO_2 diffuser:**
   - o A device that breaks CO_2 into fine bubbles, allowing it to dissolve more efficiently into the water.

3. **CO_2 drop checker:**
   - o A simple device that changes color to indicate CO_2 levels in the water. This helps maintain the ideal CO_2 concentration without harming fish.

Water Heating and Circulation

1. **Heater:**
   - o Tropical freshwater tanks typically require water temperatures between 72-78°F (22-25°C). Adjustable heaters are best for keeping the water at a stable temperature, crucial for both plant and fish health.

2. **Thermometer:**
   - o Place it inside the tank to monitor temperature consistency. Digital thermometers are more accurate, but traditional stick-on thermometers work as well.

3. **Water circulation pumps:**
   - o Used to create a gentle flow of water throughout the tank, which helps distribute nutrients and CO_2 more evenly. Proper circulation ensures plants grow evenly and prevents stagnant zones where debris might accumulate.

Aquascaping Tools

1. **Aquascaping tweezers:**

o Long tweezers allow for precise planting, especially when placing smaller plants or working with delicate species. They help minimize disruption to the substrate.

2. **Aquascaping scissors:**

o Curved or straight scissors are ideal for trimming plants. Regular pruning encourages healthy growth and prevents the tank from becoming overgrown.

3. **Spatula:**

o Helps smooth and level the substrate during initial setup, creating a neat and structured layout.

4. **Aquascaping glue:**

o Used to attach plants, like moss, to rocks or driftwood, ensuring they stay in place and grow naturally over time.

Hardscape Materials

1. **Aquarium-safe rocks:**

o Seiryu Stone, Dragon Stone, and Lava Rock are popular choices for creating natural, artistic landscapes. Ensure that the rocks you choose are aquarium-safe and won't alter the water chemistry.

2. **Driftwood:**

o It adds both aesthetic appeal and serves as a surface for attaching plants like Anubias or Java ferns. Common driftwood types include spider wood and manzanita. Be sure to soak driftwood before placing it in the tank to avoid floating.

3. **Aquarium-safe decorations:**

o Choose elements that complement the aquascape's natural theme and avoid anything that may release harmful chemicals or dyes into the water.

Aquatic Plants

1. **Foreground plants:**

o These are typically small and grow close to the substrate, creating a lush carpet effect. Popular choices include dwarf hairgrass, Monte Carlo, and Glossostigma.

2. **Midground plants:**

o Slightly taller plants like Anubias, Java fern, and Cryptocoryne add volume and texture to the midsection of the tank.

3. **Background plants:**

o Tall and fast-growing plants, such as Vallisneria, Amazon sword, and Ludwigia, provide a lush backdrop that adds depth to the aquascape.

4. **Floating plants:**

o They are optional but provide natural shade, help reduce algae by competing for nutrients, and can enhance the tank's visual appeal. Examples include duckweed, frogbit, and water lettuce.

Water Treatment

1. **Dechlorinator:**

o Tap water contains chlorine and chloramines that are harmful to fish and plants. A dechlorinator neutralizes these chemicals instantly, making the water safe.

2. **Water conditioner:**

o Helps stabilize the water parameters and supports fish and plant health by neutralizing heavy metals and other harmful substances.

3. **Fertilizers:**

o Liquid fertilizers supply essential nutrients to water column-feeding plants. Root tabs are specifically for plants with deep roots.

4. **Test kits**:

o Regular testing for pH, ammonia, nitrite, nitrate, and general hardness (G.H.) is vital for maintaining water quality. A well-balanced tank helps plants thrive and prevents fish stress.

Maintenance Tools

1. **Algae scraper:**

o A tool used to clean algae off the glass. Magnetic algae scrapers make cleaning easy, especially for larger tanks.

2. **Gravel vacuum:**

o A siphon is used for water changes and cleaning detritus from the substrate without disturbing the plants.

3. **Water change system:**

o Python water changers simplify the process of draining and refilling tanks, making maintenance quicker and easier.

4. **Aquarium safe bucket:**

o Always use a separate bucket for water changes and tank cleaning. Using one that's dedicated solely to the aquarium ensures no harmful chemicals or residues are introduced.

Aeration (Optional)

1. **Air pump:**

o An air pump helps increase oxygen levels in the water, especially in low-circulation tanks. It can also help in tanks with high bioloads or if the CO_2 levels get too high.

2. **Air stone:**
    - Creates fine bubbles that distribute oxygen evenly and increase surface agitation.

Fish and Fauna

1. **Tropical freshwater fish:**
    - Choose species that are compatible with the plants and each other. Peaceful species like tetras, rasboras, and gouramis are popular. Avoid species that may dig up plants or disturb the substrate.

2. **Shrimp or snails:**
    - Amano shrimp, cherry shrimp, and nerite snails are excellent for natural algae control and cleanup.

With these supplies, you'll have everything you need to build and maintain a vibrant freshwater aquascape, fostering both plant and animal life in a beautifully balanced environment.

Recommended Books, Blogs, and Forums

Here's a comprehensive list of recommended resources for aquascaping, including books, blogs, and forums, to help you expand your knowledge and inspiration:

Recommended Aquascaping Books

- **"The Art of Aquascaping" by George Farmer:** This book is a go-to for beginners and experienced aquascapers alike. It provides a step-by-step guide to creating stunning aquascapes, covering plant care, hardscaping, and lighting.
- **"Nature Aquarium World" Series by Takashi Amano:** A classic series of books by the legendary aquascaper Takashi Amano. His photography and philosophy on nature-inspired aquascaping revolutionized the hobby. These books are rich in inspiration and high-quality visuals.
- **"The 101 Best Aquarium Plants" by Mary E. Sweeney:** A handy guide for selecting plants for your aquascape, with detailed descriptions of care requirements, growth patterns, and aesthetics.
- **"Aquascaping:** Aquarium Landscaping Like a Pro" by Aqua Eden: This book is perfect for those looking to dive into advanced aquascaping techniques. It covers layout designs, plant care, and creating balanced ecosystems.
- **"Aquascaping:** A Step-by-Step Guide to Planting, Styling, and Maintenance" by Oliver Knott: Oliver Knott, a renowned aquascaper, offers practical advice on how to design, plant, and maintain aquascapes with an emphasis on long-term sustainability.
- **"Aquarium Designs Inspired by Nature" by Peter Hiscock:** This book provides stunning aquarium designs inspired by natural landscapes. It covers various styles, from biotope setups to more artistic interpretations of nature.
- **"Aquarium Plants:** The Practical Guide" by Pablo Tepoot: A detailed guide focused on aquarium plants, offering care tips, planting techniques, and plant identification, perfect for aquascapers aiming to master plant use.
- **"Sunken Gardens:** A Step-by-Step Guide to Planting Freshwater Aquariums" by Karen A. Randall: A thorough guide on setting up planted aquariums, covering substrates, fertilizers, plant species, and layout ideas, with a focus on low-tech and high-tech

aquascapes.

- **"The Optimum Aquarium" by Kaspar Horst and Horst Kipper:** A classic aquascaping book, highly recommended for those looking into planted tanks with a focus on balanced ecosystems and proper nutrient management.
- **"The Simple Guide to Planted Aquariums" by Terry Ann Barber and Rhonda Wilson:** A beginner-friendly book offering practical steps to starting and maintaining a planted aquarium. It's ideal for those new to aquascaping.
- **"Aquarium Plant Paradise" by Takashi Amano:** Another iconic work by Takashi Amano, this book focuses on combining aquatic plants with natural aesthetics to create serene underwater landscapes.
- **"The New Saltwater Aquarium Guide" by Albert B. Ulrich III:** While focused on saltwater tanks, it offers insights into creating thriving, visually stunning aquatic environments, with many principles applicable to freshwater aquascaping.
- **"The Complete Aquarium Logbook" by Vincent B. Hargreaves:** This guidebook doubles as a log for aquascapers, allowing you to track the progress of your tank's ecosystem while offering advice on aquascaping and tank maintenance.
- **"The Inspired Aquarium" by Jeff Senske and Mike Senske:** This book provides inspiration from renowned aquascapers around the world, featuring design techniques for creating various styles of aquascapes.
- **"Freshwater Aquariums for Dummies" by Maddy Hargrove and Mic Hargrove:** A comprehensive beginner's guide, perfect for those just starting out. It covers the basics of setting up and maintaining a freshwater tank with an emphasis on aquascaping principles.

Popular Aquascaping Blogs

- **The Green Machine Blog (thegreenmachineonline.com):** This blog, run by James Findley and his team, offers a wealth of articles on aquascaping techniques, tank setups, plant care, and product reviews.
- **ScapeFu (scapefu.com):** ScapeFu is a blog and podcast dedicated to aquascaping, with topics ranging from layout techniques to interviews with aquascaping professionals.
- **Aquascaping Love (aquascapinglove.com):** Aquascaping Love offers a mix of tutorials, product reviews, plant guides, and inspirational galleries. It's designed for all levels of aquascapers.
- **Practical Fishkeeping (practicalfishkeeping.co.uk):** Though more general in focus, Practical Fishkeeping often features aquascaping articles and tutorials that are valuable to enthusiasts looking to improve their skills.
- **Advanced Aquarist (advancedaquarist.com):** A site for aquarium hobbyists that features expert articles on aquascaping, plant care, and marine life. Their in-depth articles make it a valuable resource for advanced aquascapers.
- **Aquatic Eden (aquatic-eden.com):** A blog that focuses on the beauty of planted aquariums and aquascaping. It offers DIY guides, aquascape inspiration, and plant care tips.
- **The Aqua Design Amano Blog (adana.co.jp/en/contents):** From the creators of the Nature Aquarium style, this blog features updates and tutorials and showcases the latest trends in aquascaping.
- **The Planted Tank Blog (theplantedtank.co.uk):** A UK-based blog that provides detailed guides on planted aquariums, with articles on aquascaping, plant selection, and maintenance.

- **Aqua Forest Aquarium Blog (aquaforestaquarium.com/blogs/news):** A blog focused on nature aquariums, providing practical advice on tank setups, aquascaping tips, and equipment reviews.
- **Aquaristmagazine (aquaristmagazine.com):** A blog and online magazine dedicated to the aquarium hobby, covering a wide range of topics, including aquascaping techniques and interviews with aquascapers.
- **Shrimp Now (shrimpnow.com):** While focused on shrimp tanks, this blog offers excellent aquascaping advice for creating planted tanks suitable for shrimp.
- **Aquaticlog (aquaticlog.com):** A digital tool and blog that helps aquarists log tank parameters while also providing aquascaping tips and inspiration.
- **Fish Lore (fishlore.com):** Though focused on general fish care, Fish Lore includes a section dedicated to planted tanks and aquascaping tips, which is ideal for beginners.
- **Tropica (tropica.com/en/inspiration/blog):** Tropica's blog offers professional insights into aquascaping, including plant selection, design principles, and tank maintenance techniques.
- **The Fish Tank Blog (fishtanksetups.com/blog):** A blog with in-depth tutorials on tank setups, aquascaping techniques, and plant care aimed at beginners and intermediate aquascapers.

Active Aquascaping Forums

- **U.K. Aquatic Plant Society (ukaps.org/forum):** One of the most active forums for aquascapers, UKAPS has a wealth of user-generated content, including aquascaping journals, advice on plant care, and discussions on tank setups.
- **The Planted Tank (plantedtank.net/forums):** A huge community of aquascapers and planted tank enthusiasts. This forum covers all aspects of aquascaping, with sections for beginner advice, aquascaping layouts, lighting, and $CO2$ systems.
- **Aquarium Advice Forum (aquariumadvice.com/forums):** A general aquarium forum with a dedicated section for planted tanks and aquascaping. It's a great place to get advice, share experiences, and learn from others.
- **AquaScaping World Forum (aquascapingworld.com/forum):** This forum focuses solely on aquascaping and features dedicated sections for plant species, aquascaping techniques, and equipment. It also has a showcase gallery where members display their work.
- **Tropical Fish Forums (tropicalfishforums.com):** While it caters to all aspects of fishkeeping, the aquascaping section is full of useful tips, layout ideas, and troubleshooting advice for planted tanks.
- **Aqua Rebell Forum (aquasabi.com/forum):** A German-based aquascaping forum with international participation. It offers detailed discussions on tank setups, aquascaping tips, and plant care.
- **Fishkeeping Forum (fishkeepingforum.co.uk):** A forum for general fishkeeping that has a dedicated section for aquascaping, offering advice and project sharing among hobbyists.
- **Planted Aquarium Central Forum (plantedtankcentral.com):** A focused forum that covers all aspects of planted tanks and aquascaping, from lighting and $CO2$ to aquascaping competitions.
- **Fishforums.net (fishforums.net):** One of the largest aquarium forums, it includes a vibrant aquascaping community that discusses various styles, techniques, and challenges.

- **The Aquascaper Forum (theaquascaper.org):** A community of aquascapers sharing tips and techniques and showcasing their work. It's known for its friendly, supportive atmosphere for beginners.
- **Aquarium Substrate and Plant Forum (substrateforum.com):** A niche forum dedicated to substrate choices and planted tanks. It's perfect for detailed discussions on the foundation of aquascapes.
- **The Shrimp Spot Forum (shrimp-spot.com):** While mainly focused on shrimp keeping, this forum has a large section on planted tanks and aquascaping suitable for shrimp environments.
- **Aquatic Plant Central Forum (aquaticplantcentral.com/forum):** A long-standing forum focused on aquatic plants, aquascaping, and natural aquarium setups. It offers detailed advice on creating lush planted tanks.
- **Aquaristik Forum (aquaristik.de/forum):** A German forum with a large international presence, covering everything from planted tanks to aquascaping competitions and plant species identification.
- **Reef Central Forum (reefcentral.com/forums):** While mainly focused on reef tanks, Reef Central has a dedicated planted tank and a very active aquascaping section.

These resources will help you deepen your understanding of aquascaping and connect with other hobbyists. Happy aquascaping!

Checklist for Starting Your First Aquascape

Creating an aquascape can be an exciting and rewarding project, but it requires careful planning and execution. Here's a step-by-step checklist to guide you through the process of starting your first aquascape.

Research and Planning

- **Research Aquascaping Styles:** Explore different styles like:
  - **Iwagumi:** A minimalist, Japanese style focused on stone arrangements.
  - **Dutch:** A plant-heavy layout with a focus on color, contrast, and plant diversity.
  - **Nature:** Inspired by natural landscapes like mountains, forests, and valleys, often using a mix of wood and stones.
  - **Jungle:** Dense and chaotic, mimicking a jungle with minimal maintenance required.
- **Decide on a Theme:** Decide whether you want your aquascape to reflect a natural environment (e.g., a riverbank or rainforest) or an artistic, abstract layout. Think about the type of aquatic life and plants that will thrive in the layout.
- **Consider the Aquarium Location:** Choose a spot for your aquarium that avoids direct sunlight to prevent unwanted algae growth. Make sure there's enough space around the tank for maintenance.
- **Sketch Your Layout:** Use paper, digital tools, or 3D modeling software to visualize the placement of rocks, wood, plants, and other elements. This step is crucial to creating a well-balanced and harmonious design.

Aquarium and Equipment

- **Choose the Right Tank Size:** For beginners, a 20-30 gallon tank is ideal as it offers ample space for plants and fish while being easier to maintain stable water parameters than smaller tanks. Larger tanks give more flexibility with design but require more upkeep.
- **Lighting:** LED lights are energy-efficient and provide adjustable intensity, which is essential for plant growth and algae control. Invest in good lighting that suits your plants' needs. Different plants have different light requirements:
  - **Low-light plants:** Java Fern, Anubias.
  - **Medium-light plants:** Cryptocorynes, Amazon Swords.
  - **High-light plants:** Carpet plants like Monte Carlo and Rotala.
- **Filtration System:** Ensure your filter provides good water circulation without disturbing your aquascape's layout.
  - **Canister filters:** Excellent for larger tanks as they offer more space for biological filtration media and are generally quieter.
  - **Hang-on-back (HOB) filters:** Easier to install and maintain for smaller tanks but may not offer as much filtration power for larger, heavily stocked tanks.
- **CO2 System:** While not mandatory for all aquascapes, adding a CO2 injection system enhances plant growth dramatically, especially for more demanding species. Pressurized CO2 systems are more efficient than DIY systems.
- **Heater:** Ensure your tank's temperature suits the plants and fish you've chosen. Most tropical plants thrive in temperatures between 72-82°F (22-28°C). A quality heater with an adjustable thermostat helps maintain stable water temperature.
- **Substrate:** Choose a nutrient-rich substrate to promote plant growth.

Options include:

  - **Aquasoil (e.g., ADA Amazonia):** Provides essential nutrients for plant roots and helps maintain water acidity.
  - **Gravel:** Can be used for low-maintenance aquascapes or combined with root tabs for added nutrients.
  - **Sand:** It is best for aesthetic purposes or for creating fine details but lacks nutrients for plants on its own.
- **Aquascaping Tools:** Invest in quality tools like:
  - **Long scissors:** For trimming plants.
  - **Tweezers:** For precise planting of delicate plants.
  - **Spatula:** To smooth or shape the substrate.

Hardscape Materials

- **Rocks:** Choose aesthetically appealing stones like:
  - **Seiryu stone:** Known for its rugged look and blueish hue.
  - **Dragon stone:** Rich in textures and grooves, great for creating caves and shelter for fish.

Always check if the stones affect water pH or hardness.

- **Wood:** Choose natural driftwood like:
  - **Spider wood:** Thin and branch-like, perfect for creating natural underwater tree structures.
  - **Mopani or Malaysian driftwood:** Heavier is ideal for anchor points or central features.
- Placement: Use the golden ratio or rule of thirds when placing your hardscape. Avoid symmetry to create a more natural and dynamic layout. Use rocks and wood to create height and depth, making the aquascape appear larger than it is.

Plants and Livestock

- **Select Suitable Plants:** Choose plants based on the lighting, CO2, and nutrient levels your setup can provide.
  - **Foreground plants:** Ideal for creating carpets. Examples include Dwarf Hairgrass, Monte Carlo, and Glossostigma.
  - **Midground plants:** Help bridge the gap between foreground and background. Java Fern, Cryptocoryne, and Anubias are great choices.
  - **Background plants:** Taller plants like Vallisneria, Amazon Sword, and Ludwigia provide height and can create a sense of depth.
- **Fish and Invertebrates**: When selecting fish, consider species that won't uproot plants or disturb the layout. Suitable species include:
  - **Schooling fish:** Neon tetras, Harlequin Rasboras.
  - **Bottom dwellers:** Corydoras catfish, which help clean the substrate.
  - **Algae eaters:** Amano shrimp, Otocinclus catfish, and Nerite snails help manage algae.

Setting Up and Cycling the Tank

- **Clean the Tank and Hardscape:** Rinse the aquarium and hardscape thoroughly to remove any contaminants or dust that may interfere with water clarity or plant health.
- **Install the Substrate:** Begin by layering the substrate. A sloping layout with more substrate in the back creates a sense of depth. You can also add nutrient-rich additives under the substrate to enhance plant growth.
- **Arrange the Hardscape:** Place your rocks and driftwood first to set the foundation for your layout. Play around with the placement until it feels balanced. Use zip ties or aquarium-safe glue to secure wood to rocks if needed.
- **Planting:** Begin planting by positioning foreground plants first, followed

by midground and background plants. Use tweezers to place delicate plants without disturbing the substrate.

- **Filling the Tank:** Slowly fill the tank with water using a plate or plastic bag to avoid displacing the substrate or plants.

Cycling: Allow your tank to run for 2-4 weeks without fish to build up beneficial bacteria that help break down waste. Monitor ammonia, nitrite, and nitrate levels during this time. Consider adding liquid ammonia or fish food to kickstart the process.

Maintenance Essentials

- **Water Changes:** Conduct weekly water changes (25-50%) to maintain water quality, reduce algae growth, and remove toxins.
- **Fertilization:** Even with nutrient-rich substrates, liquid fertilizers or root tabs may be necessary, especially for fast-growing plants or high-light setups.
- **Monitor Water Parameters:**
  - **pH:** Most plants and fish prefer slightly acidic to neutral water (6.5-7.5).
  - Ammonia, Nitrite, Nitrate: Ensure ammonia and nitrite levels are zero before introducing fish. Nitrate levels should remain below 20 ppm for optimal plant and fish health.
- **Algae Control:** Algae is inevitable in most aquascapes. Control it by:
  - Reducing light intensity or photoperiod (8-10 hours a day).
  - Adding algae-eating species (Amano shrimp, Nerite snails).
  - Dosing liquid carbon (e.g., Seachem Excel) to inhibit algae growth.

Final Touches

- **Adding Livestock:** Once the tank has fully cycled (ammonia and nitrite levels are zero), introduce fish and shrimp gradually to avoid overloading the biological filter. Acclimate them slowly to the tank's water conditions.
- **Trimming and Pruning:** Regularly trim fast-growing plants to maintain the desired shape and prevent them from overshadowing smaller species. This also encourages new growth.
- **Monitor Lighting and CO2:** Adjust lighting and CO2 levels as plants grow and mature. Over time, you may need to increase light intensity or CO2 to sustain dense plant growth.

By following this checklist, you'll be well on your way to creating a thriving, beautiful aquascape!

Thank You for Your Support!

Your support means everything to me, and I hope 'The Aquascaping Master Collection' has inspired you to create stunning and thriving freshwater aquariums.

If you found this guide helpful, would you consider sharing your experience in a review on Amazon?

Your feedback not only helps others discover the art and science of aquascaping but also allows me to continue providing resources to support your journey in creating breathtaking aquatic habitats!

Simply scan the QR code below to leave your review.

Thank you for being part of this creative and vibrant underwater adventure!

Made in United States
North Haven, CT
05 November 2026

21747511R00070